FORWARD DRIVE

SIERRA CLUB BOOKS ▪ SAN FRANCISCO

FORWARD DRIVE

THE RACE TO BUILD "CLEAN" CARS FOR THE FUTURE

JIM MOTAVALLI

The Sierra Club, founded in 1892 by John Muir, has devoted itself to the study and protection of the earth's scenic and ecological resources—mountains, wetlands, woodlands, wild shores and rivers, deserts and plains. The publishing program of the Sierra Club offers books to the public as a nonprofit educational service in the hope that they may enlarge the public's understanding of the Club's basic concerns. The point of view expressed in each book, however, does not necessarily represent that of the Club. The Sierra Club has some sixty chapters coast to coast, in Canada, Hawaii, and Alaska. For information about how you may participate in its programs to preserve wilderness and the quality of life, please address inquiries to Sierra Club, 85 Second Street, San Francisco, CA 94105.

www.sierraclub.org/books

Published by Sierra Club Books, in conjunction with Random House, Inc.

Library of Congress Cataloging-in-Publication Data

Motavalli, Jim.

Forward drive: the race to build "clean" cars for the future / by Jim Motavalli.

p. cm.

Includes bibliographical references and index.

ISBN 1-57805-035-9 (alk. paper)

1. Hybrid electric cars. 2. Fuel cells. I. Title.

TL221. 15.M68 2000 629.22'93—dc21 99-32154

10 9 8 7 6 5 4 3 2 1

BOOK DESIGN BY BARBARA M. BACHMAN

FOR MARY ANN, AND FOR

MAYA AND DELIA,

IN THE HOPE THEY'LL

GET TO DRIVE

"CLEAN CARS"

"THE FUTURE IS, WELL, HERE."

—Honda ad for the company's
new hybrid vehicle in *The New York Times*,
January 7, 1999

CONTENTS

FOREWORD

I WAS BORN IN the Los Angeles Basin. By the time I was twenty, in 1970, I had spent most of my life inhaling the dirtiest air in the nation. I finally decided to take action on a particularly smoggy day when I was having trouble breathing. I was also having trouble seeing an intersection a block away. I was fed up.

Being a resourceful young man, I opened up the yellow pages and searched for a car dealer with a difference. I called the number and a man named Dutch answered. "Electric vehicle sales and service, may I help you?" he said.

Thus began a journey that continues to this day. From the simple Taylor-Dunn purchased from Dutch in 1970 to the sophisticated EV1 I drive today, it has been an interesting trip.

I believe that the days of the internal-combustion engine are numbered and that advanced-technology "clean cars" will gradually own the lion's share of the market. A power source, like the gasoline engine, that dissipates 80 percent of its energy before it ever reaches the rear axle is too wasteful in the modern era and must be replaced by something more efficient and less polluting.

In this comprehensive work on the future of the automobile, Jim Motavalli talks to all the key players, from consumer advocate Ralph Nader and top-auto-executive-turned-electric-bicycle-convert Lee Iacocca to "hypercar" pioneer and energy researcher Amory Lovins.

Though each perspective is given fair treatment, there is nothing wishy-washy or noncommittal about this work. With years of experience both writing an auto column and reporting on environmental issues (two worlds that rarely come together), Motavalli has excellent credentials for sifting through the mountain of information on the latest auto technologies. And he makes highly educated guesses about what probably will happen, and what probably won't.

He feels, as I do, that continuing our near-total dependence on foreign oil, with all the national security concerns that come with it, is unwise. He also concludes, quite rightly, that we can't continue to burn dirty fuels in polluted cities such as Los Angeles and Houston, which have air too filthy to meet minimal federal standards.

But Motavalli is not living in a dream world of the way things ought to be. This is not a simple good hat/bad hat story, and the happy ending hasn't yet been written. The American public has not embraced the electric automobile in any great numbers, largely because battery cars can't deliver the range and performance consumers have come to expect. And although natural-gas pipelines nearly cover this nation, where is the infrastructure to make new car buyers want to drive a natural-gas vehicle off the showroom floor? Fuel cells have incredible potential, but when will they be light enough and cheap enough to form the basis for an affordable car? Hybrid cars, while not a zero-emissions solution, have the important advantage of using our existing network of gas stations.

This book delves deep into questions like these and avoids

simplistic answers. I will tell you this, though: Reading *Forward Drive*, I learned more about the past, present, and future of the automobile than I ever have from all my voracious reading on the subject.

The personal computer and the cell phone are examples of the advanced technology that keeps getting better and cheaper every day. At long last, technological wizardry is being applied to the automobile, with the result being a new generation of cleaner cars. The gasoline engine represents yesterday's solution to our transportation problems. Your next car might look, sound, and perform totally differently from the vehicle you drive today.

And wouldn't that be nice?

Ed Begley Jr.
Los Angeles

PREFACE

THE DAY I GOT my driver's license, I drove the family station wagon two hundred miles into the next state, so ecstatic was I at the prospect of liberation from the parental orbit. I bought a car as soon as I could, which was almost immediately. I tinkered with it on weekends. I subscribed to car magazines. I was a car nut.

In some ways, I still am. I write a syndicated auto column and have become accustomed to the shiny new test cars in my driveway. I've been to exotic places to drive new models, and gone on television and radio as an automotive guru. The car has certainly been good to me, but I'm becoming disenchanted.

Driving down the empty, sun-dappled country lanes pictured in car commercials is certainly fun, but inching to work on crowded asphalt isn't. As I sit in traffic, windows tightly rolled up against the toxic fumes from all those idling exhausts, it's hard to remember the freedom I once felt. Through my work as an environmental reporter, I've learned a great deal about the harm caused by fossil fuels as they're extracted, transported, burned, and fought over. And that's affected my love affair, too. I'm not sure you can call me a car "enthusiast" anymore, though I'm still hopelessly addicted to them.

During the past few years, as I looked at the trends—not only are new car registrations skyrocketing while fuel economy stagnates, but the average driver's yearly mileage is on the upswing, too—I found it's not hard to imagine World War III being fought over the last few gallons of Middle East oil. But before despair grabbed hold, I began to hear about some new technologies that offered, if not a way out of auto addiction, at least an alternative to tailpipe asphyxiation, fossil-fuel dependence, and the swift onset of global warming.

The information I came across was often buried deep inside technical journals, or couched in the auto industry's insider language. But it described a personal transportation revolution that was becoming tantalizingly close. The alternative propulsion systems I heard about weren't exactly new—the history stretched back 160 years—but rapid technical advances were making them practical for the first time. I began to see the need for a book that could explain these new developments to readers who'd somehow failed to get their engineering Ph.D.s. To understand what was happening, people needed more than schematic diagrams. Context was missing. Could the auto industry break away from its symbiotic relationship with Big Oil and embrace something startlingly different? And if these radical cars were built, would people buy them?

The more closely I looked at what was going on in the research laboratories, the more excited I got. Driven by air pollution legislation, the threat of global warming, and a suddenly animated international competition, carmakers were making quantum leaps forward in technology. Here, at last, were vehicles that promised to not only greatly reduce pollution but also to perform better, be more reliable, cruise farther, and last much longer than anything the public had ever seen. This could be, in short, a whole new evolution of the automobile, at a time when such progress was desperately needed.

The cars I write about in this book won't end gridlock. Only public transportation can do that, but the prospects for a rail and bus renaissance aren't great. America has poured $329 billion into new highway construction in the last forty years without raising much public indignation, but the halls of Congress are full of cries for the elimination of Amtrak subsidies, which have cost us all of $20 billion since 1971.

Unfortunately, even if we doubled public transportation ridership, it wouldn't help that much. In his book *Future Drive,* author Daniel Sperling notes sadly that such a Herculean effort would lessen vehicle trips in the U.S. by only a few percentage points. We're that dependent on private cars.

Environmentalists don't love cars, and they shouldn't. The appalling cost of these "insolent chariots" in their first century is fully detailed here. But as America sprawls ever farther out from the city centers, where public transit works best, we're only adding to our auto addiction.

"The car will not vanish, so we must clean it up," writes Hank Dittmar of the Surface Transportation Policy Project. This book is about "clean cars," a hopeful turn in the ultimately rather depressing history of the automobile. Carmakers are delivering cars powered by high-efficiency hybrid drives (with both conventional internal-combustion power *and* electric motors) and emission-free fuel cells that run on hydrogen. What's more, fuel cells offer the tantalizing possibility of ushering in an entirely new energy economy. They may soon power our watches, our laptop computers, and our homes.

Far from being a limited resource, hydrogen is the most available element in the universe, constituting 80 percent of all matter. The sun burns hydrogen, but this, the lightest element, is found in pure form almost nowhere on earth. It has to be separated from other substances, such as natural gas or methanol. And therein lies the problem now consuming platoons of engineers.

If hydrogen is produced by renewable energy sources, such as photovoltaics or geothermal power, it can be a perfect zero-emission loop, with drinkable water the only by-product. The fuel-cell car can be an electric vehicle with none of the drawbacks of batteries. "It's like a dream, isn't it?" one auto company fuel-cell expert told me, and he's right. The promise of fuel cells can realistically be compared in importance to Thomas Edison's development of a workable incandescent light. Progress has been rapid. Fuel-cell cars are on the road right now, and they could be in mass production as early as 2004. On two consecutive days just before Earth Day 1999, General Motors announced a major fuel-cell partnership with Toyota, and DaimlerChrysler unveiled the California Fuel Cell Partnership, which will put fifty of these high-technology cars on the state's roads by 2003.

Ironically, fuel cells and hybrid cars are evolving at a time when not only is public confidence in short-range battery-powered electric vehicles (EVs) at a low ebb but cheap fuel prices have seemingly made the whole discussion moot. To paraphrase the average driver of a sport-utility vehicle, "Who cares about gas mileage when prices at the pump are so low?"

Today's battery EVs are, indeed, very limited, with few able to cover more than ninety miles between time-consuming recharges. Batteries *are* getting better, but not quickly enough to satisfy consumers. Until they can offer the 250-mile range most consumers have come to expect, it's hard to see battery cars breaking out of niche status. In contrast, a lightweight fuel-cell car could travel one thousand miles or more between pit stops.

And the era of cheap, abundant gasoline is likely to be short. A new era of taxation and increasingly stringent regulation, together with diminishing returns in oil exploration, is likely to soon increase the cost of America's joyride many times over.

The first chapter of *Forward Drive* looks at today's alternative-fueled cars as part of a historical continuum that stretches back to a child's toy built for a Chinese emperor in the seventeenth century. While there have been striking successes, such as the impressive market share earned by battery cars in the early part of this century, the EV time line mainly shows tantalizing near misses and brilliant failures, all of them swept away by the growing hegemony of the internal-combustion engine after 1915.

Chapter 2 examines the devastating environmental consequences of the gasoline automobile's one-hundred-year career, a catastrophe that's going global as rising incomes in relatively car-free countries like China persuade billions of people to leave public transportation and their bicycles behind.

But can the Third World learn from the West's mistakes and give fossil-fuel dependence a miss? The technologies introduced in chapter 3 have that potential, though many pitfalls and a great number of technical hurdles remain. Chapter 4 gets into the passenger seat with the pioneers who've put their principles into action by buying and leasing EVs. While the EV hasn't yet penetrated Middle America, it certainly has its passionate fans, including nuts-and-bolts techies, gear-obsessed "early adopters," and a virtual colony of Hollywood environmentalists.

Chapter 5 is a visit to the "skunkworks," the small labs and think tanks that domestic automakers and independent companies have set up to bring the new cars into the consumer mainstream. Many of the U.S. carmakers remain conflicted, lobbying against clean-air reforms from one office while pushing EVs from another. But there's a growing financial commitment to the new technologies, as well as a strong desire not to be left behind by the well-funded research under way in Europe and Japan, covered in detail in chapter 6.

Far from existing in a vacuum, the clean car's world launch is buffeted by a variety of political and financial forces, as well as by the advocacy work of a few key players and the wavering interest of the international press. There are visionaries who predict a clean break from the past, and special-interest groups trying to make it all go away. Chapter 7 gets caught up in these swirls and eddies.

The federal government is itself a player in the automotive debate, through an ambitious partnership with industry aimed at creating fuel-efficient prototypes. Chapter 8 examines the recent history of automotive environmental regulation and explains why it isn't more of a force today. The evolving consensus has more to do with international automotive competition and state regulations (especially California's) than it does with federal directives. Finally, chapter 9 puts the new EVs into context as a vital element in the broader picture of sustainable transportation. And because both the technology and the political climate are subject to change, it looks at the different ways clean cars could develop in the next decade.

I wrote part of this book in a pastoral paradise named Point Reyes Station, in California, where the traffic woes of the rest of the world seemed far away. No one can remember gridlock in town, unless you count the slight snarls caused by families coming back from the beaches on weekend afternoons. There are a lot of bikes leaning against lampposts in town, and a lot of people out walking.

But they get the San Francisco papers in Point Reyes Station, and while I was there they were full of bad news about unhealthy air days, tangled traffic loops, oil spills, and the scandals shaking up the Muni light-rail system, once a model for the rest of the country. Clean cars won't solve all of these problems. Even in a zero-emission vehicle, we'll still sit in traffic, though because there's no smoking exhaust the wait will be more pleasant. If transportation is to move efficiently in the new millennium, we'll have to combine improvements in the personal automobile with a wide array of other reforms, including moratoriums on suburban sprawl, construction of new in-town housing, and development of an interconnected rapid-transit network.

At the end of the twentieth century, fuel-efficient and hydrogen-powered cars can seem like the answer to a question nobody's asking. But the auto industry is, for once, looking ahead, and seeing not only the end of the oil era but also a global-warming crisis that won't easily be solved without changing the way the world drives. The automakers certainly aren't green, but their new cars represent a giant leap forward in the movement toward truly sustainable transportation.

Fairfield, Connecticut
November 1999

ACKNOWLEDGMENTS

MANY PEOPLE HELPED me keep all four wheels on the ground through this complex story. From inside the auto industry, I'd like to thank Fred Heiler at Mercedes-Benz, Dick Thompson and Mark Leddy at General Motors, Mark Amstock, Wade Hoyt, and Dave Hermance at Toyota, Ben Knight at Honda, Jeannine Fallon at Volvo, and Brendan Prebo at Ford. Debby Roman of Ballard Power Systems was very helpful, as were James Worden and Karl Thidemann of Solectria. Princeton's Joan Ogden, Jim Cannon of Energy Futures, and Sandy Thomas of Directed Technologies explained how a hydrogen economy would work. Ralph Nader, Hazel Henderson, and Amory Lovins helped me see both how this technology could change the world, and the obstacles that might get in the way. The Mesa Refuge in Point Reyes Station, California, was open to me at a critical juncture, and I was able to write a big chunk of the book on a laptop there, looking out over Tomales Bay. For that, I thank Peter Barnes, Pam Carr, and Ann Dowley. My fellow residents, Jeanne Trombly and Bill Shireman, dispensed Peet's Coffee and excellent suggestions. The Institutes for

Journalism and Natural Resources made me a Fellow just when I needed to be one, and for that I am indebted to Frank Allen and Andrew Weegar. Doug Moss, Steve Sawicki, and Tracey Rembert kept *E Magazine* going, against daunting odds, while I was on the road. For getting me interested in writing books in the first place, I thank Hans and Kate Koning. For hand-holding, sage advice, a sharp editorial eye, and great patience I offer the most sincere gratitude to Sabine Hrechdakian, my agent. David Rothenberg, who introduced me to Sabine, offered strong encouragement and the example of his own prodigious multimedia works. At Sierra Club Books, Helen Sweetland has become an ally, friend, and helpful motivator. For giving an unruly manuscript a coherent shape and for providing far more helpful counsel than most writers can expect these days, I'm deeply indebted to my editor, Linda Gunnarson. Friends who helped along the way include Robert Davey, Elizabeth Hilts (whose "inner bitch" provided a useful reference point), and Ron Williams. Finally, great torrents of thanks and IOUs are due to my wife, Mary Ann, for making it possible for me to be peripatetic for a summer, and to my daughters, Maya and Delia, for putting up with a daddy who wanted to be an author.

FORWARD DRIVE

CHAPTER 1

PULLING THE PLUG: A BRIEF HISTORY OF ALTERNATIVE MOTION

I KNEW ELECTRIC AND steam cars had a colorful past, but I didn't know just how colorful until I delved into the literature from the earliest days of motoring. Moving people from one place to another has never been an easy business, and the florid commentaries and brightly colored illustrations that began appearing as early as the eighteenth century reveal all the political passions and social conventions of their time. Steam cars, driven by gout-ridden dandies, upset the English peasantry's horses. Manure-shoveling "dirt boys" were put out of work in New York. Women driving horseless carriages shocked the proper Bostonians. I quickly concluded that there's more to our evolving transportation history than a simple recitation of technological advances.

The misfortunes of long-ago automotive inventors, geniuses, and madmen, struggling (usually in vain) to interest a fickle public and overcome vested interests, offer a mirror for our own troubled times. It's never simple to buck tradition, even when that tradition is destructive to all concerned. Looking back from our omniscient vantage point, we might think that our ancestors would have been glad to see all those troublesome equines put out to pasture. But today's cars won't go gently to history's junkyards, either.

Most people think of the automobile as the invention that ushered out the nineteenth century and welcomed in the twentieth, with a blast of exhaust smoke. It was in 1894, for instance, that Frank and Charles Duryea of Springfield, Massachusetts, took their first orders for a gasoline buggy, giving tentative birth to a new industry that would come to influence everything from urban planning to social life.[1]

But the very earliest cars, a strange and wonderful mélange of the practical and the eccentric, were built hundreds of years before the Duryea brothers took their first test drives down Springfield's Maple Street, and they ran on steam. That the world heaved a massive sigh of indifference at their uncertain and halting appearance on the quagmires then known as roads speaks volumes to our present predicament, when the thought of being *without* automobiles is impossible to contemplate.

Leonardo da Vinci thought about carriages that could move under their own power in the fifteenth century and left drawings showing rudimentary steering and transmission systems. In 1510, the German Renaissance painter Albrecht Dürer sketched a complex and undoubtedly very heavy royal carriage, propelled by muscle power through geared cranks. The two yeomen pictured in Dürer's drawing would have had to be stouthearted indeed to get the richly adorned vehicle moving.

THE AGE OF STEAM

It's a long way from fanciful etchings to working cars, but those were not long in coming. Little is known of the Flemish Jesuit priest and astronomer Ferdinand Verbiest, but he is reliably believed not only to have mapped the boundary lines between Russia and China but also to have created a miniature four-wheeled steam carriage for the Chinese emperor Khang Hsi in the period between 1665 and 1680.[2] The opinions of the emperor and his courtiers about this undeniably advanced, self-propelled machine are not known, but Verbiest left detailed plans of his two-foot-long unmanned "car," and working models have been constructed from them. Was this steamer the first automobile? Maybe. At least until the first practical internal-combustion engine was developed in 1860, constructing a workable steam car for the road became the personal obsession of any number of scientific geniuses and eccentrics, very few of whom received anything but ribald laughter and scorn for their trouble.

Automotive historians can only imagine the scene in Paris in 1769 when the Frenchman Nicholas Cugnot, a distinguished military engineer in the service of Empress Maria Theresa of Austria, retired from active duty and began working, under royal commission, on his idea for a steam-powered military truck. The finished motor carriage may have been capable of only six miles per hour, but it moved. This, the world's first automotive test drive, sufficiently loosened the royal purse strings to fund a second and larger model. (Such bottom-line requirements have influenced automotive testing ever since.)

Cugnot's second steam carriage still exists in a French museum, and it's a strange sight indeed. It has front-wheel drive, for one thing, with the boiler hanging off the nose, creating such an unbalanced weight distribution that the vehicle was barely steerable. This perilous situation was probably responsible for what became the world's first motor accident. According to some accounts, the carriage hit a wall and rolled over, landing its designer in jail.[3] Small wonder, then, that the carriage never turned a wheel again.

The French, who invented the bicycle and then let the fruits of their invention slip away to America and England, did the same thing with the automobile. After Cugnot, the scene shifted to England, where steam carriages were quickly adopted into public transportation, which, were roads better, might have become a considerable force. The significance of the technical breakthroughs was matched only by the public's apathy toward them. Who could imagine that these outlandish contraptions, easily outrun by even the lamest horse, would ever become practical?

The English pioneer James Watt, whose technical innovations made the Age of Steam possible, applied for and was granted a patent for a steam carriage in 1786, but was wise enough not to actually build it. Watt was so afraid of explosions from high-pressure boilers that he had a covenant written into the lease for any potential tenants of his home, Heathfield Hall, stipulating that "no steam carriage shall on any pretext be allowed to approach the house."[4]

Richard Trevithick, a pioneer of the high-pressure steam engine, developed and patented a locomotive-like carriage with a boiler and smokestack that attained a heady nine miles per hour on Christmas Eve 1801. Trevithick's huge car, which had eight-foot rear wheels, made several relatively trouble-free trips, though during one the vehicle went awry and tore out some garden railing. The London Steam Carriage "proved to be a financial disappointment."[5]

Another intrepid British inventor, Goldsworthy Gurney of the Surrey Institute, built a long-distance steam car that, in 1825, made an eighty-five-mile round-trip journey without incident in ten hours. His coach was later damaged by anti-machinery Luddites, who, according to a contemporary account, "considered all machinery directly injurious to their interests . . . [and] set upon the carriage and its occupants, seriously injuring Mr. Gurney and his assistant engineer, who had to be taken to Bath in an unconscious condition."[6]

American steam pioneers faced ridicule and censure, too. No one remembers Oliver Evans today, but this unassuming and notably unsuccessful early-nineteenth-century engineer built the first self-propelled vehicle in America—and it also *swam.* This unsung mechanical genius also constructed the first high-pressure boiler in the U.S., and created an automation system for a grain mill that prefigured Henry Ford by 150 years.[7]

Evans, a Philadelphian, built his twenty-ton Orukter Amphibolos in 1805 as a dredger to excavate the city's waterfront. To get it to the Schuylkill River, a mile from his workshop, he drove it up Market Street at four miles per hour, attracting crowds. The twenty-five cents he charged onlookers was the only money Evans ever made from steam vehicles. The Amphibolos was notably unsuccessful as a dredger. It was sold for scrap, and Evans's later plan to build a fleet of produce-carrying trucks was rejected by the Lancaster Turnpike Company, which concluded that they'd probably shake to pieces on the terrible roads. They were undoubtedly right, and besides, America, like England, was developing a fast and efficient rail system. By 1850, there were nine thousand miles of railroads in the U.S., and nothing but muddy horse tracks for other traffic between most major towns and cities.

Given that, it's remarkable that an unbowed Evans predicted in 1812, "The time will come when people will travel in stages moved by steam engines, from one city to another, almost as fast as birds fly. . . . A carriage will set out from Washington in the morning, the passengers will breakfast in Baltimore, dine in Philadelphia and sup in New York the same day."[8]

The inventors struggled to make their fire-belching vehicles practical, and an English entrepreneur named Walter Hancock was the first to offer regular passenger routes of the type Evans imagined. Between 1824 and 1836, Hancock had nine steam coaches taking on paying customers. One, sure to terrify passengers with the nickname *Autopsy,* traveled between London and Islington, hauling three omnibuses, a stagecoach, and fifty passengers.[9]

But Hancock's revolutionary bus service didn't attract enough brave souls; it lasted only five months before failing financially. It probably would be going too far to say a conspiracy killed the steam coaches, but certainly the powerful and established railroad interests threw some spikes in the road. As Ken Purdy reports in his classic book *Kings of the Road:*

> The railroads of the day took a very dim view indeed of the kind of competition steam coaches obviously could offer, and arrangements to make things difficult for the upstarts were not hard to contrive. A four-shilling tollgate charge for horse-drawn coaches, for example, could easily be raised to two pounds and eight shillings for a steam coach. . . . The steam coaches had received, in 1831, a clear bill of health from the House of Commons, but even Parliament could not prevail against the will of "the interests"—not the first time it has turned out so.[10]

Even more crippling than high road taxes was the infamous "Red Flag Act," passed in 1865, which restricted self-propelled vehicles to two miles an hour in town and four in the country. In addition, a person carrying a red flag had to walk in front of anything moving under its own power.

In the next half century, steam carriages would make their mark, particularly in America. One particularly innovative model, shaped like a railroad locomotive, was commissioned by the government Indian agent Joseph Renshaw Brown to carry food and supplies to isolated bands of Sioux Indians before the Civil War, though the bad roads in the hinterlands of

Minnesota made it impractical for the purpose. More successful were the brightly painted fire engines that appeared in several cities after the war was over. Steam cars survived the onslaught of internal combustion in the first decade of the twentieth century, and marques such as Stanley, Locomobile, and White had many admirers, who loved the cars' silent operation, range, and absence of a crank handle (a necessary accessory in the days before electric starting). This latter advantage, shared by electric cars, was a prime consideration in an era when misapplied cranking could break your arm.

But Stanleys and Whites could take up to half an hour to work up "a head of steam," consumed great amounts of water and wood fuel, and scared people who read accounts of boiler explosions. When rapid technological advancements radically improved the gasoline engine after 1905, the market for steam cars dried up. By 1911, White and Locomobile abandoned steamers and switched to gasoline power.

THE ELECTRIC CAR PLUGS IN

Electrics lasted longer, though they too would eventually lose the race to the wider-ranging gasoline car. The dignified electric at least put up a pretty good fight before succumbing.

As the British authors of the book *Automania* suggest, the horrendous roads that had done much to stifle development of the early steam car actually worked to the advantage of the first electrics: "Until American roads were improved, almost all cars kept within the city limits where the short range of the electric car was no great drawback."[11]

How bad were the roads? The United States had twenty-seven thousand miles of them as early as the 1830s, but most were dirt tracks. In 1903, while both gas and electric cars were busy being born, only 10 percent of U.S. roads were paved,[12] which helps explain why the first automobiles were inevitably "high wheelers." In addition to the seas of mud that formed whenever it rained, horse traffic turned streets into cesspools. In cities like London and New York, armies of street sweepers were employed to clean up an average of forty-five pounds of dung per horse per day. One consequence of the motor age was reflected in a contemporary British postcard showing a somber "dirt boy" watching the whizzing auto traffic. "Nothing Doing" reads the caption.

It's hardly surprising, then, that the quiet, clean electric car found favor. Electric vehicles grew out of early experiments with electric trains and trolleys and became possible only with the invention of the practical storage battery in 1859. The first true electric car may well have been a three-wheeled carriage made by Magnus Volk of Brighton, England, in 1888.[13] In company with that seventeenth-century steam car built for the Chinese court, the earliest electrics also drew royal patrons; witness the four-passenger carriage, complete with a one-horsepower motor and twenty-four-cell battery, that Immisch & Company built for the Sultan of Turkey, also in 1888.[14]

Electrics made their way to America soon after. William Morrison of Des Moines, Iowa, made a stir at the World's Columbian Exhibition in Chicago in 1893 with a six-passenger electric wagon that carried full loads of delighted patrons.[15] To them, the self-propelled automobile was a fascinating novelty but a toy of no immediate practical use, since trains occupied the only usable roadbeds, and trolleys (there were 850 systems in the United States by 1895) worked the streets in cities.

Electric cars first made a discernible impact on American lives through their use as taxis, particularly in New York City. By 1898, the Electric Carriage and Wagon Company had a fleet of twelve sturdy and stylish electric cabs—with well-appointed interiors able to accommodate gentlemen in top hats—plying the city streets. As in contemporary horse carriages, the driver sat outside on a raised platform.[16]

Through small-scale, successful businesses like this, the electric car gradually won acceptance. By 1900, Americans could choose their motive power, and at the first-ever National Automobile Show, held that November in New York City, polled patrons overwhelmingly favored electric as their first choice, followed closely by steam. Gasoline ran a distant third, getting only 5 percent of the vote. There were 1,681 steam, 1,575 electric, and only 936 gasoline cars made that year.[17]

Unlike the steam car tinkerers, who were sometimes lucky to escape with their lives after demonstrating their inventions in the village square, many of the early gasoline car pioneers became industrialists whose names are well remembered today. Gottlieb Daimler, Henry Ford, Ransom Olds, Carl Benz, William Durant (who founded a little company called General Motors), James Packard, and John M. Studebaker are a few of the men (it was, alas, exclusively a men's field) who survived the vicious competition of the early years of motoring.

Many of the men who made their careers selling internal combustion had a lifelong fascination with electric cars. The German Dr. Ferdinand Porsche, for instance, built his first car, the Lohner Electric Chaise, in 1898 at the age of twenty-three. The Lohner-Porsche, as it was also known, was absolutely groundbreaking in that it was the world's first front-wheel-drive car (excluding, perhaps, Nicholas Cugnot's carriage) and also introduced such revolutionary concepts as four-wheel brakes and an automatic transmission. Its use of electric motors in the four wheel hubs was revived for the 1971 moon lander and has been brought out again in some of today's hybrids, which have both gas and electric power. (Amazingly enough, Porsche's second car was a hybrid, using an internal-combustion engine to spin a generator that provided power to the electric motors, again in the wheel hubs. On battery power alone, the car could travel thirty-eight miles).[18]

Henry Ford was also fascinated by all things electrical, and he received early encouragement from the "Wizard of Menlo Park," Thomas Edison, who became a close friend. The two met at a Brooklyn dinner party in 1896, with Edison allegedly telling the thirty-three-year-old Ford this about his gas buggies: "Young man, that's the thing, you have it. Keep at it. Electric cars must keep near to power stations. The storage battery is too heavy. Steam cars won't do, either, for they have to carry a boiler and fire. Your car is self-contained—it carries its own power plant—no fire, no boiler, no smoke, no steam."[19]

Nevertheless, neither Ford nor Edison gave up on electric cars, and in 1914, the Ford Motor Company announced it would introduce a rechargeable vehicle of its own, powered by Edison's batteries. But by then it was clear that the electric car was on the road to nowhere, and the project never went beyond a few prototypes and some inflated press releases. It was the end of an era and, unfortunately, the end of significantly funded electric-car development for almost six decades.

Women at the Wheel

Image marketing is nothing new: colorful illustrations placed early luxury cars in the driveways of swanky estates, with well-dressed "men of the world" at the wheel. For electric and, to a lesser extent, steam cars, which sold on their quiet, easy operation, the appeal was plainly to women.

Hartford, Connecticut, manufacturer Colonel Albert Pope, who had sold five hundred electric cars by 1897, famously proclaimed that "you can't get people to sit over an explosion,"[20] and that was indeed true for many early auto buyers, particularly women, who were taking to the road in ever-greater numbers. Electric cars were bought for (if not necessarily by) women, and female drivers were heavily featured in early electric advertising. The Flanders 20 Coupe was, for instance, "the ideal vehicle to carry Milady into the shopping district or on her round of social duties,"[21] according to a typical ad from 1910. The Woods Electric of 1911 was "the chosen car of men of affairs, as well as the favorite conveyance of Her Highness, the American woman."[22]

As Virginia Scharff observes in her book *Taking the Wheel: Women and the Coming of the Motor Age,* "Women were presumed to be too weak, timid and fastidious to want to drive noisy, smelly gasoline-powered cars. Thus at first, manufacturers, influenced by Victorian notions of masculinity and femininity, devised a kind of 'separate spheres' ideology about automobiles: gas cars were for men, electric cars were for women."[23] Electrics, which reached mass production soon after the new century began, roughly paralleling gasoline cars, were also built to what was assumed to be a woman's specifications. Although some economy models were plain, many others were offered with particularly opulent overstuffed interiors, bud vases, and lace curtains. They were front parlors on wheels, an effect enhanced by rows of facing seats.

If women had bought into the automakers' sexist expectations in great numbers, the early electric might have had a longer heyday. "The men who manufactured and sold cars made assumptions about what women wanted," Scharff told me in an interview. "They assumed that women wouldn't be bothered by the electric's range problem, because their 'domestic sphere' was limited in scope. But the campaign to sell electrics to women may have been primarily a marketing phenomenon. Most women who were able to drive drove the family car, and after 1907 that was usually a gas-powered Model T." Two-car households were rare, and electric cars were often not only expensive but unable to cope with the bad roads that existed everywhere but in the largest cities. Scharff cites a 1915 survey of women drivers in then-rugged Houston, Texas: Only 30 of the 425 female car owners there drove electrics.

The End of an Era

The device that largely killed the electric car as a viable product—the self-starter for internal-combustion cars—was itself first marketed to women. "Any Woman Can Start Your Car" read a 1911 ad from the Star Starter Company that featured a bonneted woman at the wheel of a crankless gas buggy. By "cranking from the seat, and not from the street," she was eliminating one of the last major advantages the early electric car had.

Charles Franklin Kettering's effective starter quickly spread throughout the industry, and sales of electrics plummeted, down to 6,000 vehicles, or 1 percent of the industry, by 1913. That same year, the Ford Model T alone sold 182,809. Electric carmakers closed or consolidated. Of the twenty-seven companies that were selling electrics at the height of the market in 1910, less than ten survived until the end of World War I. A few specialized firms, such as industry leader Detroit Electric, limped into the 1920s, catering mainly to a diehard coterie of wealthy women.[24]

One interesting product of the industry's death throes was the Woods Dual Power coupe, a resounding flop produced in Chicago from 1917 to 1918. Like Ferdinand Porsche's second car—and the ultramodern Toyota Prius—the Woods Dual Power was a hybrid, with a four-cylinder gasoline engine under the hood and an electric motor alongside it. Woods had been producing electric cars since 1899, and the company clearly saw that the end was near. The hybrid was Woods's attempt to have it both ways, but the car was relatively expensive and its advantage in fuel economy meant little in an era when gasoline was very cheap. By all accounts, the Dual Power worked well, but few were sold. One of the only surviving examples of this historically significant curiosity sits on shredded tires in the basement of the Petersen Automotive Museum in Los Angeles.[25]

Ironically, the electric car died out just as some of the major objections to it, such as high price, slow speed, and short range, were finally being addressed. The last Detroit Electrics could reach thirty-five miles per hour, which was competitive in the early 1920s. And the cheap, cheerful, and light Dey runabout of 1917 was offered to a largely indifferent public at just $985. No doubt, electric-car technology would have continued to evolve, but with no industry to support it, research and development went as flat as one of Edison's batteries after a fifty-mile run.

By 1926, there were more than eight million automobiles in America, and a brand-new coast-to-coast federal highway for them to ride on in U.S. Route 40, but the electric car (indeed, the whole idea of alternative power) had lost any mainstream commercial support. Electrics, steamers, and, eventually, hydrogen cars became the province of backyard tinkerers and post office box entrepreneurs, who sold conversion kits in the back pages of *Popular Mechanics*.

Stirrings of Revival

Electrics have made sporadic reappearances on the American landscape, mostly in response to fuel crises, like the Arab oil embargo of 1973. Unfortunately, most have been underdeveloped conversions, such as the utility-endorsed Henney Kilowatt of 1959 to 1961, which started life as a French Renault Dauphine. Some 120 Henneys were built, which is a good run for a postwar EV.

By the early 1960s, Americans tired of Detroit's befinned land yachts had demonstrated, by buying Volkswagen Beetles in the hundreds of thousands, that they were ready to consider economy cars. The Big Three responded to the foreign invasion with such "compacts" as the Ford Falcon and Dodge Dart, cars that seem laughably large and thirsty now. In Europe, tiny fifty-miles-per-gallon "microcars" like the BMW Isetta enjoyed a brief vogue. Some independent optimists thought the time might be right for a modern electric. "Are Electric Cars Coming Back?" asked *The Saturday Evening Post* in 1960.[26]

But many of the "new" EVs that appeared over the next two decades seemed predestined for failure, with bizarre pop-riveted bodies, tacky parts-bin interiors, and below-the-radar marketing. Take the Urba Sports Trimuter of 1981, a "space-age" three-wheeler with a head-frying pop-up canopy, a needle nose, and a top speed of sixty miles per hour that the enterprising owner was supposed to build himself from a set of $15 plans.[27] Other half-baked monstrosities included the Free-Way Electric, which looked like a bug and was probably as easily squashed; the three-wheeled Kesling Yare, with styling straight out of *A Clockwork Orange;* and the B&Z Electric, which in photos appears to have been made of ill-fitting scrap wood.

Florida-based Sebring-Vanguard actually managed to sell 2,200 two-seat, plastic-bodied CitiCars in the wake of the 1973 Arab oil embargo, but these phone booths on wheels, powered by golf-cart batteries and with a top speed of only thirty miles per hour (as well as a range of forty miles "in warm weather"), weren't likely to build a loyal customer base. Viable commuter EVs would have to wait until the first reliable conversions appeared in the late 1980s.

Meanwhile, other would-be auto moguls were trying to revive the steam car. Robert Paxton McCulloch, a chain-saw millionaire, lost a sizable part of his fortune on a steam prototype, the Paxton Phoenix, between 1951 and 1954. And a colorful character named William Lear, who invented not only the eight-track stereo but the first aircraft automatic pilot (as well as the business jet that bears his name), spent $15 million in 1969 dollars on a turbine bus and a 250-horsepower turbine steam-car concept. Both manufacturers succeeded in producing quiet, efficient, and modern steam engines with minimal pollution, but neither was fully developed. Even though his bus suffered from reliability problems and poor gas mileage, Lear was so confident that he tried (without success) to enter a steam car into the 1969 Indianapolis 500.[28] Even British sports-car maker Austin-Healey was working on a steam car, one with four-wheel drive no less, in 1969.

Nevertheless, even relatively wealthy entrepreneurs like McCulloch and Lear soon discovered that they lacked the resources to successfully market a competitive car, no matter how brilliant the science behind its propulsion system. Gary Levine could write, in his 1974 book *The Car Solution,* "The cry 'bring back the steam car' is now uttered with increasing frequency by those familiar with the nation's health, energy, and transportation problems,"[29] but that cry proved easy to ignore. To really launch a new, or new-old, technology would take the deep pockets and experience of an auto company. But aside from the occasional gaudily-stickered prototype to display at auto shows, draped in bikini-clad models, the car industry just wasn't interested in alternative power. It wouldn't be until air-quality regulations tightened, and engineers made technical breakthroughs with hybrid and fuel-cell cars, that its attitude would change.

CHAPTER 2

A DIZZYING RIDE: INTERNAL COMBUSTION'S RAPID RISE AND COMING DECLINE

I LOVED CARS LONG before I knew there was any reason to worry about their effect on the environment or be concerned about the smoke that poured from their tailpipes. In the 1960s, ignorance like mine was widespread in the United States, maintained by a powerful automotive lobby and a complacent federal government. Highway congestion, though already bad, was somewhat masked by an expanding national highway grid, and most people celebrated the migration to the suburbs that the new roads aided and abetted. Cars were equated with freedom, and ads of the period showed happy vacationing families riding in roomy sedans, with the uncrowded interstate stretching out in front of them.

The V-8 engine, dominant since the 1950s, and the "muscle cars" of the late 1960s and early 1970s were symbols of an impatient country on the move. The in crowd romanticized in the Beach Boys' "I Get Around" could say, "We always take my car 'cause it's never been beat." The icon of the age was the AC Cobra, a tiny and barely controllable British sports car with a huge Ford 427 engine.

I bought into all of this, subscribed to the car magazines, and put up posters of the Dodge Charger in my room. But on April 22, 1970, because my friends were going, I attended the first Earth Day celebration in Washington, D.C., and it became a turning point—for me and for many other Americans. Since then, our knowledge about the automobile's effects on the environment has certainly grown. And most of the news we've gotten is bad.

If ever a human invention has reached a critical moment in its history, it is the internal-combustion automobile, whose one hundredth anniversary was celebrated in 1996. We are literally choking to death on our enduring love affair with the gasoline-powered car. Since 1969, the U.S. vehicle population has grown six times faster than the human population, 2.5 times faster than the number of households, and double the rate of new drivers. As Matthew L. Wald put it in *The New York Times,* "They bid fair to become the dominant life form."[1] Despite being only 5 percent of the world's population, Americans own 34 percent of the planet's cars and drive an estimated two trillion miles annually. Between 1900 and 1984, we sent more than 640 million motor vehicles to the scrap heap.[2]

Even in places where the air is still breathable, traffic congestion and gridlock turn commuting into a daily endurance test. In Bangkok, Thailand, for instance, rush-hour traffic struggles to reach a fast walk, and drivers carry portable toilets with them for the inevitable emergency. In Athens, Greece, the death rate climbs 500 percent on bad-air days. In São Paulo, Brazil, dirty air and clogged streets have forced officials to set up a rotation system for drivers that keeps one-fifth of the city's cars off the road at any given time.[3] Cars are also a huge problem in Tel Aviv, Israel, where smog is predicted to reach Mexico City levels (the worst in the world, with ozone levels three times safe limits) by 2010; already, it has led to outbreaks of asthma and bronchitis in the city and in nearby Jerusalem. In Prague, Czech Republic, smog occasionally forces the police to set up roadblocks and keep all but essential traffic out of the city center. In Singapore, drivers pay a premium for licenses that allow them unlimited access to the highways.

But even as we're realizing what our continuing reliance on the private automobile is costing us, we're adding another fifty million of them to the planet's burden every year. By 2030, there could be one billion cars taking up space on the earth—an astounding figure that means, in effect, that the auto industry will produce as many cars in the next thirty years as it did in its first century. Car making is now the largest manufacturing activity on earth. Motor vehicles consume half the world's oil and account for 14 percent of its greenhouse gas emissions.

And in half the world's cities, the biggest source of air pollution is exhaust emissions.[4] The World Bank estimates that in Asia, thousands of people die prematurely every year from filthy air. The problem is most acute in China, which has a heavy reliance on coal.[5] China is also leading the ominous Third World rush to "modernize" through the use of private cars. Although almost 80 percent of its travel is now either on foot or by bicycle, the world's most populous and rapidly industrializing country had auto sales of an estimated 1.6 million in 2000, and could have 100 million cars by 2015. Bicycle use there is being discouraged. According to the journal *Geophysical Research Letters,* if 400 million Chinese drivers hit the road in cars over the next fifty years, the plume of tailpipe exhaust would "bathe the entire western Pacific in ozone," extending all the way to the United States.[6]

The *Baltimore Sun* reported in 1996 that China's leaders are committed to building "an automobile culture . . . that one day may well rival America's." There were less than two million private cars in China that year, but the numbers are rising steadily. The U.S. auto industry sees rich opportunities. In 1997, GM joined with Shanghai Automotive Industry Corporation to build Buick Regals and Centuries in China, to the protest of some Chinese critics, who said the cars were too big. Other U.S. manufacturers want a piece of the pie. Wayne Booker, of Ford's international operations staff, says the Chinese market is "vitally important to the long-term success of Ford." Chrysler recently unveiled its two-cylinder, fifty-miles-per-gallon Composite Concept Vehicle for the China market; its prototype is at least made of lightweight plastics (partially reclaimed from soda bottles) and meets contemporary emission standards.[7]

If China's drivers end up behind the wheel of Chryslers and Buicks, it's because of the global reach of the American auto giants, which, because of international mergers, are becoming less American. But the companies started out small, with their modest factories concentrated by an accident of history in and around the city of Detroit, Michigan.

AN INDUSTRY GROWS UP

The United States had only eight thousand registered cars and trucks in 1900 and almost 200 million in 1995. That incredible growth was due initially to the work of hundreds of companies, but it eventually became the exclusive province of what became known as the Big Three: Ford, General Motors, and Chrysler. To understand how the industry grew up, you have to imagine the incredible entrepreneurial excitement existing in Detroit at the turn of the last century, all inspired by the prospect of private automobiles and roads to run them on.

Some of the first cars, the Duryeas and Popes, were actually built in New England, which had been a center of the wagon maker's art. But it was Detroit that caught car fever—and Detroit manufacturers that chose to specialize in this new and highly speculative endeavor. In Detroit alone, thirty-eight auto companies were formed in 1900, forty-seven in 1901, and fifty-seven in 1903. But many of the companies put together in haste failed just as quickly: twenty-seven in 1903 and thirty-seven in 1904. "The atmosphere of the town was reminiscent of the Gold Rush," write Peter Collier and David Horowitz in their book *The Fords: An American Epic,* "a fury of automotive wildcatting based on the certain knowledge that fortunes would be made and lost in the next few years."[8]

Probably the most daring of these wildcatters was the founder of General Motors, a man enormously influential in his day who nevertheless died in obscurity and is now little remembered. His name was William Crapo Durant, and he came from a prominent Flint family. (His grandfather was the governor of Michigan.) Durant began as a highly successful wagon maker, so cars were a natural next step. If Henry Ford invented the automotive assembly line, then Durant deserves credit for the automotive conglomerate. Acquisition and merger were in his blood.

The GM story begins with a young inventor named David Buick, who, in 1903, convinced Durant that horseless carriages would replace horse-drawn wagons. Buick, whose background was in building bathtubs, loved to tinker with engines, but he wasn't much of a businessman. In 1904, its first year of operation, Buick Motor Company sold only thirty-seven cars. Durant became involved with Buick within three months of its launch, and was soon its guiding force. The earliest years were difficult, but as onetime employee Walter Chrysler reported, Durant "could charm a bird right down out of the tree."[9] By 1907, Buick was the second largest carmaker in the country. Durant saw, when others did not, that there would soon be demand for a million or more cars a year and, with Buick as his base, he went on a buying spree after incorporating General Motors in 1908.

Oldsmobile, Cadillac (which already had its reputation for high quality), and Oakland were all in the fold by 1909, and the company kept on growing, acquiring such valuable franchises as Champion Spark Plug, Delco, and Fisher Body in a fury of vertical integration. By 1910, GM controlled almost a quarter of the U.S. auto industry.

But Durant also bought a whole batch of industrial lemons, and they soured his reputation. It was in 1910 that GM's board, complaining that Durant couldn't integrate the towering edifice he'd created, ousted him, only to watch as he acquired and energized the fledgling Chevrolet and then used it to gain control of GM a second time. Durant pioneered the auto air conditioner and founded the General Motors Acceptance Corporation so people could buy cars on credit, but he couldn't stop buying ill-fated companies, and it was all over for him by 1921. Durant stayed in the car business as the head of Durant Motors, but that company filed for bankruptcy in 1933. His later years were spent as the proprietor of a bowling alley, which, he promised all who would listen, would be the first of a chain.[10]

One of the few companies that Durant never succeeded in buying was Ford, whose chief executive was far too independent-minded to consider submitting himself to outside management. Although he'd been tinkering with horseless carriages for years (and ran his first working gasoline engine in the kitchen of his Dearborn home in 1893), Henry Ford was actually a latecomer to automobile manufacturing. By the time the Ford Motor Company was set up in 1903, Oldsmobile, Cadillac, and Buick were well-established businesses. But Ford, a Dearborn farm boy who was also a self-taught mechanical genius, soon established his own niche by building cars that were both simple to operate and affordable. Instead of raising the price slightly each year, as most companies did, Ford's Model T got *cheaper*, dropping from $950 in 1909 to an incredible

$240 for what was a much more advanced car in 1925.[11] Ford's farm background undoubtedly helped him keep perspective about his customers' needs in a still largely rural America. The approach worked: by 1918, nearly half of the cars in the world were Model Ts, and Ford's efficient assembly lines were capable of turning out ten thousand of them a day.

The auto industry was a small, incestuous community in the early years of the new century. Ford bought his first engines and chassis from the hard-drinking Dodge brothers, John and Horace, who were one of his principal early backers. The self-made former railroad machinist Walter Chrysler, who lent his name to the perpetual also-ran of the Big Three (a term he himself originated), had suffered under William Durant's high-handedness while helming Buick and was a General Motors retiree by the time he formed the Chrysler Corporation in 1925. The company's first car was a sensation when, because there was no space at the auto show across town, it was unveiled in the rented lobby of New York's Commodore Hotel. The overnight popularity was vital since, without attracting new backers, Chrysler would have lacked the capital to put it into production. With his namesake established, Chrysler expanded by buying Dodge two years later. Like Ford, Chrysler focused on inexpensive cars for the masses, and as a tribute he brought the third Plymouth off the line to Dearborn and presented it to Henry Ford and his son, Edsel.[12] As the industry's founding fathers died in the 1940s and 1950s, however, the close ties between the Big Three automakers broke down and the rivalries between them became less personal.

THE ROOTS OF AUTO ADDICTION

In just the twenty years between 1900 and 1920, the U.S. developed a mechanized auto industry capable of delivering inexpensive cars to the masses. But production capacity alone couldn't put the country on wheels. How did we become so dependent on the automobile, and so quickly? There were many factors, a lot of them deliberate social policy. The Federal Aid Roads Act of 1916 encouraged coast-to-coast construction of paved roads, usually financed by gasoline taxes (a symbiotic relationship if ever there was one). After 1939, with a push from President Franklin Roosevelt, limited-access interstates began to make rural areas accessible. Among the first passengers on the superhighways were returning World War II veterans, who were financing their new suburban dream homes with federal loans. They didn't call the movement away from the cities "sprawl" back then, but that's what it was.

There wasn't necessarily anything sinister about all this. Highways were seen by many as just one aspect of the technological progress that would make life easier for all. In his book *1939: The Lost World of the Fair,* David Gelernter argues that the General Motors Futurama exhibit, which took fair-goers through the imagined world of 1960, complete with a fourteen-lane Express Motorway that would crisscross the nation at one hundred miles per hour (with car spacing controlled by "radio beams"), was wildly popular precisely because of the freedom and mobility the interstate highways promised.

Some modern historians, Gelernter says, suggest "that the Futurama exhibit was the launchpad of an evil GM scheme to foist highways on an unwilling public—and that is absurd."[13] At the same time, however, there were vigorous protests against new highways in many cities, precisely because some people could see beyond the glitter to the roads' ultimate impact on neighborhoods and urban life in general. As early as 1898, a *Life* magazine cartoon entitled "A Sunny Day in 1910" depicted a city scene dense with smoke from the bumper-to-bumper traffic.

But the switch to cars and away from public transit wasn't happening fast enough for some corporate leaders. In 1922, when only one in ten Americans owned a car, General Motors launched a covert campaign to destroy the then-dominant U.S. public transportation systems.[14] The campaign, which took thirty years to fully implement, focused on the country's clean (powered by electricity), convenient (trolleys on most avenues ran every few minutes), and safe (accidents were infrequent) streetcar system.

GM, in partnership with Standard Oil and Firestone, began by buying the largest bus maker in the U.S. It then secretly funded a company called National City Lines, which by 1946 controlled streetcar operations in eighty cities. Despite public opinion polls that, in Los Angeles for instance, showed 88 percent of the public favoring expansion of the rail lines after World War II, NCL systematically closed its streetcar systems down until, by 1955, only a few remained. A federal antitrust investigation resulted in both indictment and conspiracy convictions for GM executives, but obliterating a public transportation network that would cost hundreds of billions of dollars to reproduce today cost the company only $5,000 in fines.[15]

Did GM kill the streetcars all by itself? Obviously not. There were many competing forces, and municipal sentiment in some areas did favor buses over trolleys. Writing in *Transportation Quarterly,* Cliff Slater argues that the streetcar lobby was itself a powerful force that delayed what might have been an even earlier takeover by buses, which were faster and offered more frequent service.[16] But GM's intentions—and its overt actions—were clear, and its conviction stands as a milestone public rebuke.

Destroying the rail lines and replacing them with buses was only the first step. If private cars were going to dominate American transportation, they needed new roads to run on. GM also stands behind the creation in the early 1930s of the National Highway Users Conference, otherwise known as the highway lobby, which became the most powerful pressure group in Washington. GM promotional films from the immediate post-war years proclaim interstate highways to be the realization of "the American dream of freedom on wheels." Ford, meanwhile, busily promoted the "Two-Ford Family."

GM President Charles Wilson, who became secretary of defense in 1953, used his position to proclaim that a new road system was vital to U.S. security needs. He was assisted by newly appointed Federal Highway Administrator Francis DuPont, whose family was then the largest GM shareholder. Acting on a bill introduced by Senator Albert Gore Sr., Congress approved the $25 billion Interstate Highway Act of 1956. "The greatest public works program in the history of the world," as Secretary of Commerce Sinclair Weeks called it, was on, and with it were planted the seeds of our current gridlock.

As the highways expanded, they carried Americans farther and farther from the city. Architects such as Frank Lloyd Wright, who designed a suburban community in 1935, saw the interstates as encouraging a new, more graceful form of single-family living. Today, fifty years after ground was struck for Levittown, the influential planned community on Long Island, the process it heralded has become known as sprawl, a seemingly endless stretch of minimalls and housing developments, reached almost exclusively by private cars. Since the mid-1950s, for instance, the city of Phoenix, Arizona, has grown from seventeen to well over four hundred square miles, and its traffic tie-ups are nightmarish. Between 1970 and 1990, Greater Chicago swallowed up 46 percent more land area, though its population increased by only 4 percent in that time. Similarly, New York City and its environs grew by 61 percent from 1965 to 1990, but added only 5 percent more people.

EARLY WARNING SIGNS: LOS ANGELES IN CRISIS

The first cracks in the federal highway monolith appeared in California, a bellwether state for many American trends. Fast-growing Los Angeles, with its cloverleafs sending suburb-bound commuters every which way, was soon the poster city for highway congestion and, later, for smog as well.

It's not surprising that California was the first state to get tough on auto emissions, because auto exhaust now accounts for 90 percent of the state's carbon monoxide, 77 percent of its nitrous oxides, and 55 percent of its reactive organic gases, according to the California Air Resources Board. On some days, ozone levels can be three times the federal limit. In recent years, California's air *has* gotten cleaner. (There was only one Stage 1 smog alert in 1997.) The progress is largely the result of the country's most stringent regulations, which, in some cases, have had carmakers building special pollution-controlled "California editions" of their cars. The basin city of Los Angeles was still winning smoggiest-city-in-the-nation honors as late as 1998, but on most days now you can at least see the Hollywood sign from downtown.[17]

It's worth looking at Los Angeles' auto addiction in some depth, because its air-quality crisis, and the increasingly draconian solutions that followed, are both very recent. Not that there weren't signs and prophecies. The Portuguese explorer Juan Rodríguez Cabrillo, who saw the plain that was to become Los Angeles in 1542, was so struck by the haze produced by Indian campfires that he called the place Bay of Smokes.

The Spanish settlers did their part to transform the landscape, importing cattle by the hundreds of thousands for their vast ranchos and destroying the delicate native grasslands in the process. From its mission roots, characterized by vast individual landholding, a subdivided city grew. In his second book about Los Angeles, *Ecology of Fear,* Mike Davis shows clearly how what had been a bucolic agricultural region in the nineteenth century, made famous by the orderly rows of orange trees seen on color postcards, was rapidly replaced by a form of runaway development that consumed the natural landscape and left little open space and few public parks in its wake.

Between 1907 and 1927, Los Angeles added two million people, most of them lured by the healthy climate and unparalleled scenery. They also came to find work in the booming oil industry; southern California floated on a sea of fossil fuel, and by 1909 the state was supplying 80 percent of American fuel oil. The black gold needed consumers:

> The fledgling public relations industry got to work selling the automobile to folks . . . who were otherwise happy with the streetcar. Newspapers, movies and *Sunset* [magazine] ballyhooed the private car both for its convenience and as the ideal means of escape from growing city cares. Top down and filled with carefree revelers, automobiles posed against backdrops of . . . the ubiquitous orange groves and snow-capped mountain ranges of southern California.[18]

Land speculation was particularly fierce during the 1920s building boom, which was unfettered by any public planning. The urban design firm Olmstead Brothers, of which renowned park planner Frederick Law Olmstead Jr. was a partner, warned in a 1930 report to local authorities that the developing city would likely disappoint its new arrivals:

> The beaches, which are pictured in the magazines to attract eastern visitors, are suffering from the rapid encroachment of private use; the wild canyons are fast

> being subjected to subdivision and cheek-by-jowl cabin construction; the forests suffer annually from devastating fires; the roadsides are more and more disfigured by sign boards, shacks, garages, filling stations, destruction of trees.[19]

The first recorded smog event in the city didn't occur until 1943, when California was rapidly adding both population and cars because of the ready availability of jobs in weapons factories. Not coincidentally, Los Angeles also decommissioned its last trolley around this same time.

Smog, a combination of industrial smoke and fog, was named by London doctor H. A. Des Voeux in 1911, when coal, not auto exhaust, was the major component. Thirty years later, cars were still above suspicion in Los Angeles. People suffering from smarting eyes, difficulty breathing, nausea, and vomiting thought they'd been hit with a gas attack and blamed it on a chemical leak from a nearby factory.

The city's Bureau of Smoke Control was established in 1945, as the war was ending and Los Angeles' legendary sprawl was beginning. Studies of the growing smog problem convinced officials that it needed to be attacked on the county and state levels. By 1947, Governor Earl Warren had created air pollution control districts throughout California.[20]

Meanwhile, the populations of both people and cars continued to swell. In 1930, California had six million people and two million cars. By 1950, both figures had doubled, and the car population was growing faster than the human one.

In 1950, Dr. Arie Haagen-Smit (a scientist later appointed chairman of the California Air Resources Board by Ronald Reagan) was the first to discover that when the emission of man-made nitrogen oxides and hydrocarbons, at least half of it from cars, is combined with ultraviolet light from the sun the result is smog. (Dr. Haagen-Smit would also warn—in 1969—that air pollution could cause a "greenhouse effect.") But the headlong pace of postwar growth continued, with tract housing spreading into the San Fernando Valley and, through the Interstate Highway Act of 1956, new roads springing up to service the new commuters. Around this time, California state highway patrolmen started wearing gas masks because of poor air quality on the freeways. "Daytime dimouts," during which high noon looked like late afternoon, became commonplace.

In 1962, a Harvard University study said that carbon monoxide poisoning "is becoming increasingly serious because of the increased density of smog and the concentration of idling vehicles in the metropolitan areas. Small amounts of carbon monoxide are absorbed rapidly by the bloodstream."[21] But even though the academic studies were piling up, Los Angeles got no relief from a state slow to impose air-quality reforms. The cars on California roads in the early 1960s had no pollution-control equipment of any kind. As the United States was celebrating the first Earth Day, April 22, 1970, California was choking, with an ozone (a main ingredient of smog) concentration five times the national standard. By then, there were twelve million cars in the state, and the newer ones were required to have rudimentary pollution-control devices. But the state government's biggest "assault" on the problem was telling homeowners not to burn leaves.

What put Los Angeles' air on the road to recovery? It was a combination of factors. In 1963, American cars got their first mandated air-pollution device, the simple positive crankcase ventilation (PVC) valve, which recycles unburned blowby gases. The first federal Clean Air Act, which empowered the government to set air-quality standards, was enacted that same year, though it took until 1969 to get actual pollution targets in place. And then there was OPEC. The 1973 oil embargo forced

Detroit to think small. As cars like the Pinto and Chevette hit the market, average fuel economy skyrocketed and pollution took a slight breather. Between 1970 and 1989, the average fuel economy of American cars got dramatically better, from 13.5 miles per gallon to 20.3. In 1975, Los Angeles' ozone concentration was .39 parts per million, significantly lower than the .59 parts recorded in 1970. (Still, there were 118 dangerous Stage 1 smog days in 1975.)[22]

Despite auto company protests, reducing tailpipe emissions has not been astronomically expensive. At a cost of only about $300 each for additional equipment, by the late 1990s California's cars had the cleanest internal-combustion engines in the world. The state has also become the nation's incubator for the alternative-fuel automobile, with CALSTART, a state-funded nonprofit consortium, helping start-up EV businesses take root and achieve what it calls "critical mass." Not surprisingly, Los Angeles is at the center of all this new activity. Southern California Edison operates one of the nation's biggest electric-car fleets, and electric Specialty buses make runs from Los Angeles International Airport to the long-term parking lots. Some of the city's omnipresent used-car lots have been transformed into electric-conversion shops. To fully understand the city's enthusiasm for EVs, you have only to visit it on a bad-air day, when the natural glories of the San Gabriel Mountains are lost in an ominous brown mist.

THE CAMPAIGN AGAINST EVS: CARMAKERS LOBBY CALIFORNIA'S REGULATORS

By the late 1980s, California's Air Resources Board (CARB) was wielding power over the auto industry far out of proportion to its size. Who would have thought that automakers, the unquestioned kings of the road since the turn of the century, would be put back on the path by a tiny state agency with a minuscule budget and a staff made up of what are politely called "policy wonks"?

Much of what CARB does involves the issuance of highly technical emission standards of interest to only a few power plant managers and emissions professionals, but in 1990 it threw down a gauntlet to an astonished auto industry. For the right to sell cars in California's rich market, it said, the major companies would have to start producing emission-free vehicles (only battery electric cars fit the definition) or face stiff fines. CARB's mandates specified that 2 percent of the automakers' fleets would have to be zero-emission vehicles (ZEVs) by 1998, a figure that climbed to 10 percent in 2003.

The auto industry reacted as if rear-ended by a semi. The mandates intruded on their cherished belief that the market, not regulations, should decide what kind of cars people should buy. And they believed, at least then, that people wouldn't buy EVs in any form. To their great frustration, CARB waved away their warnings that the technology wasn't ready. After several years of quiet pressure, hoping that the regulators would see reason, the industry, aided by its even better funded allies in the oil business, decided on a very public fight.

The lobbying phase of the industry campaign, which ran through most of 1995, had several arms. Californians Against Hidden Taxes (CAHT) was mostly funded by the Western States Petroleum Association (WSPA), a trade group representing oil companies in California, Arizona, Nevada, Washington, Hawaii, and Oregon whose members include Mobil, Shell, Chevron, and Arco. Working closely with CAHT was another organization funded by WSPA, Californians Against Utility Company Abuse (whose acronym, opponents gleefully point out, is CAUCA). CAHT and CAUCA were joined at the hip. CAHT's chief executive officer, Doug Henderson, was also, according to state filings, CAUCA's chief financial officer. Both groups listed the same address in Burlingame, which is the home base of public relations and political consultants Woodward & McDowell. CAUCA's news bureau director, Barbara Simpson, also worked out of the Woodward & McDowell office.[23]

CAHT's executive director was Anita Mangels (known as "the Mangler" by her foes in the environmental community). An abrasive but brilliant spokeswoman for her cause, Mangels had the daunting duty of getting a plurality of Californians to change their minds about EVs. In its own documents, the industry groups had to admit that the public liked EVs, at least in theory. The American Automobile Manufacturers Association (AAMA), which worked alongside but not directly with CAHT and CAUCA, tried to keep that fact confidential. An internal AAMA memo from 1995 says, "Recent surveys indicate a majority of Californians believe . . . ZEVs [zero-emission vehicles] . . . are a 'workable and practical' means of reducing air pollution. This is a shift from surveys and focus group results of 1993, and may indicate greater consumer acceptance of electric vehicles."[24] In other words, people would buy them, with or without mandates.

The groups made their point with billboards, newspaper and magazine ads, and plenty of public appearances on California talk radio. In one such appearance, Mangels was the guest of *Hot Talk* host Lee Rogers on KSFO-AM in San Francisco. Rogers didn't need much convincing. "I don't know how long it's going to take for the rest of the country to catch up to what the enviro-Nazis are trying to do to us," he bellowed, "but I expect some folks are going to wake up when they find out that the price of their new car is going to go up to subsidize a few rich idiots who want to buy electric cars that have been forced on the market by those crazies at the California Air Resources Board."

Mangels couldn't have agreed more, and she knew just what buttons to push, citing *Tonight Show* host (and car buff) Jay Leno as one of those "rich idiots" about to be subsidized by the citizenry to buy an electric car. "I just wonder if Mr. Leno realizes . . . that the taxpayers are going to be putting that $7,500 [subsidy] in his pocket," she said, "not the shareholders of General Motors, not anybody but the taxpayers."[25]

The cash-rich auto and oil companies—acting together or not—made a formidable lobbying team, their materials preying mostly on fear of new taxes and higher car prices. A WSPA "fact sheet" distributed in 1995 estimated that electric-car mandates would add $2,000 to the cost of every conventional car on California's roads, thus reducing new car sales in the state "by 10 percent or more."[26] A CAHT petition from the same period upped the ante, stating, "We oppose the assessment of nearly $18 billion in hidden taxes and other costs to promote electric, natural gas and other alternative-fueled vehicles."

Environmentalists counter that the oil companies themselves enjoy billions in taxpayer subsidies. The Union of Concerned Scientists, in a 1995 report entitled "Money Down the Pipeline: Uncovering the Hidden Subsidies to the Oil Industry," claimed that in 1991 the industry received $2 billion in corporate income tax benefits from the federal government, and $4 billion in state and local tax breaks.

WSPA, which consistently ranks among the top five lobbyist employers in California, declines to say how much it poured into the campaign against the EV mandates. AAMA ran its own campaign. It placed classified ads for an agency that could "create a climate in which the state's mandate requiring automakers to produce a fixed percentage of electric vehicles beginning in 1998 can be repealed." AAMA found its contractor in the Los Angeles–based public relations firm Cerrell Associates, and it did indeed lead to climactic change.

In the six months ending in November 1995, Cerrell President Hal Dash says, the auto companies spent $500,000 on its campaign (by itself dwarfing the $160,000-a-year budget of the opposition California Electric Transportation Coalition). AAMA's work was educational, according to Dash. "There are no front groups, no astroturf lobbying," he said. ("Astroturf" is the derisive term environmentalists apply to organizations that masquerade as green.)

In their testimony at CARB hearings, auto executives reinforced the lobbying campaign with strongly worded doubts about EV performance. Ford's John Wallace, who would later captain the electric Ranger program, was nonetheless pessimistic about them in public. "As anybody who is familiar with today's battery technology will tell you, EVs are not ready for prime time," he said. GM's Robert Purcell Jr., later to helm the GM hybrid and fuel-cell projects, added, "We need to view EVs in business, not regulatory, terms."[27]

Melanie Savage of CALSTART, the quasi-state consortium that lobbies for electric-car business in California, described the anti-EV effort as "a very extensive multi-million-dollar campaign against the mandate that effectively undermined faith in the technology. For every dime we spent, they spent a dollar. And they used incredible tactics. They bused in old people from Orange County to the CARB public hearings, gave them box lunches, and told them that their tax bills would go through the roof so that a small number of people could have electric cars."

Cecile Martin of the California Electric Transportation Coalition added that "CAHT was on all the far-right talk shows, fanning the flames. They used advanced electronics to have automatic dial-ups to legislators' offices. They dragged the process out for a year and a half by bringing in their high-priced attorneys and making request after request for data. They really riled people up." One of the best things the anti-EV campaign did, Martin says, was hire Anita Mangels. "She's excellent at what she does, the mistress of the sound bite." Savage adds, "At the hearings she'd be at the door, nailing the press as they came in. But you have to ask people who employ tactics the way Anita does, 'Why are you so afraid of people having a choice?' "

In an interview, Mangels claimed it was choice that she was after. "We're not opposed to the development of technology; we're opposed to government dictating what technology has to be developed and sold, and to the taxpayers picking up that cost," she said. "All such products should live or die in the private sector." She dismissed the idea that her coalition spent its way to a sea change in public attitudes, maintaining that CAHT was, in fact, a grassroots group.

Without even having to directly subsidize them, the industry also benefited from the encouraging words of its supporters in the automotive press. *Car and Driver* called CARB "the most environmentally draconian government agency in the nation" and speculated that "some invisible toxin [is] circulating in the climate-control systems of government offices that interferes with the neurotransmitters in the human cerebral cortex responsible for comprehension of things automotive." In *The Boston Globe,* car columnist John White said "everything short of M&Ms" had been tried to power EVs, "and still we do not have an electric car capable of winter operation, cheap enough for someone with less income than the Sheikh of Bahrain and with room enough for the whole family."[28]

CARB got the message. While it was undoubtedly influenced by other factors, such as the very real problems developing a suitable EV battery, CARB caved in after it became apparent that the antimandate lobbying was having a strong effect on public opinion. Polls that had shown nothing but goodwill toward EVs now revealed an undercurrent of worry—about subsidies, about battery technology, about range. In

March 1996, CARB backed down on its requirement that 2 percent of an automaker's fleet be ZEVs by 1998, but kept the mandate calling for 10 percent by 2003.

The industry was victorious. "Whatever small part we played in rescinding the 1998 mandate, we're pleased," said Gerald A. Esper, of AAMA's Detroit bureau. "We believe that the decision by CARB to rescind its mandates was the correct thing for them to do, the only logical thing to do given the information that was available."

Influencing CARB was expensive. Between 1991 and 1995, according to a study entitled "Pollution Politics" by the California Public Interest Research Group (CALPIRG), oil companies and automakers spent nearly $34 million to influence public policy in the state, the sum including $29 million in lobbying, $3.97 million in donations to statewide and legislative candidates, and $945,000 specifically to the gubernatorial campaign of Republican Governor Pete Wilson. WSPA's lobbying expenditures were $7,349,718 in the period, the CALPIRG study shows.[29]

CARB's Jerry Martin says the coalition ran "a very negative campaign, took a lot of quotes out of context, and used ridiculous numbers and tried to make them real," but he denies that pressure from the business coalition—or any other external factor—caused it to change its six-year commitment to a 1998 mandate. And although its eleven members—five elected officials, five technical experts, and a full-time chairman—are all appointed by the governor, Martin denies that then Governor Pete Wilson played any role in the new policy. His influence? "None," says Martin. "It's certainly worth mentioning that we work for the governor, but there was no pressure. Not to my knowledge."

Was Wilson, who then had presidential ambitions and had been publicly warned to repeal the mandates by Governor John Engler of Michigan, really a passive observer? It's worth noting that the oil industry contributed $866,613 to his campaigns between 1991 and 1995 (Arco was his single biggest supporter in the period, weighing in with $369,950). *The Sacramento Bee* quoted an unidentified CARB spokesman in late 1995 as saying that Wilson's office had instructed the agency to draft a repeal of the mandate, but CARB officially denied the charge. In a subsequent interview, however, Environmental Protection Agency head Carol Browner seemed to assume that the decision to back off was basically Wilson's. "I was surprised and disappointed by Governor Wilson's actions in suspending the mandates," she said. "They were very helpful in encouraging electric-car development."

While WSPA's motives are transparently clear—the oil industry stands to lose trillions of dollars if EVs succeed and Americans stop consuming 13.3 billion gallons of gasoline a year—industry groups like the now-disbanded AAMA (whose budget, once $20 million, jumped to $34 million by 1995) represented a constituency that could be seen to have divided loyalties. The irony of GM contributing to the fight against the 1998 ZEV mandate was that it directly affected the marketability of a product it spent $300 million developing, the EV1, which did indeed meet a skeptical public when it was first offered, for lease only, in California and Arizona in 1996.

REINVENTING REALITY: THE AUTO COMPANIES' HISTORY OF FIGHTING FEDERAL REGULATION

The CARB mandates were hardly the first time the auto industry had balked at government mandates and ignored pollution warnings. It was just the first time the battle was on the state level. There is a long and particularly inglorious history to carmaker combativeness, with the federal government losing skirmish after skirmish.

In 1926, lead (under the code name "ethyl") was approved as a gasoline additive, despite copious expert medical testimony to its hazards at a U.S. Public Health Service hearing. Carrying more weight were the words of Dr. Emery Hayhurst, a paid consultant to the Ethyl Corporation, who testified that leaded gasoline was being used with "complete safety and satisfaction." From then until 1989, when Congress finally outlawed leaded gas, 15.4 billion pounds of lead dust were spewed into the air by automobiles, resulting in the average person having as much as one thousand times more lead in their bodies than did pre-Columbian Indians.[30]

Ralph Nader's 1965 landmark book *Unsafe at Any Speed,* while justifiably celebrated for its pioneering treatment of auto safety, also closely examined the growing body of pollution science and accompanying industry reluctance. As late as 1953, Nader reported, even as the Los Angeles Air Pollution Control District had definitively targeted the auto industry, Ford's Dan J. Chabek assured a concerned customer that air pollution and car exhaust had no relationship. "The Ford engineering staff, although mindful that automobile engines produce exhaust gases, feels that these waste vapors are dissipated in the atmosphere quickly and do not present an air pollution problem," Chabek wrote. "Therefore our research department has not conducted any experimental work aimed at totally eliminating these gases. The fine automotive power plants which modern-day engineers design do not 'smoke.' "[31]

In 1964, the British-born Hazel Henderson, now a distinguished futurist and critic of the global economy but then a New York homemaker, formed a grassroots group called Citizens for Clean Air. Publicizing the then-novel idea that New York's air pollution was caused by power plants (65 percent) and cars (35 percent), Henderson barnstormed the radio and TV talk shows. Hugh Downs's *Today Show* gave her nine minutes. "I went on and said that oil was far too precious to burn, that we should be driving electric and fuel-cell cars," she says. "These ideas were being discussed even back then." Citizens for Clean Air gave the Ford Motor Company an award for unveiling an electric car, but the auto industry wasn't serious about actually producing it.

The emissions-cutting "blowby" crankcase ventilation system, an extremely simple but effective device, was finally introduced for Detroit's 1963 model year after a threat of federal legislation by then Health, Education, and Welfare Secretary Abraham Ribicoff. But even then, not all cars were so equipped.[32] A year later, the industry used delaying tactics to try to stall the introduction of pollution-control devices, claiming they could not possibly be ready until the 1967 model year. But after a California law required them by 1966, the industry met the deadline without difficulty. By 1969, Chrysler chief engineer Charles M. Heinen was telling the Society of Automotive Engineers that "the main battle against automotive air pollution has been won." He opined that further expenditures on cleaning up the air weren't worth it "from a social, scientific, medical and economic standpoint."[33]

In 1971, speaking candidly but caught for posterity on President Nixon's secret recording system, then Ford Motor Company President Lee Iacocca complained bitterly about the planned introduction of shoulder belts and headrests, calling them "complete wastes of money." He added, "We are on a downhill slide the likes of which we have never seen in our business. And the Japs are in the wings ready to eat us up alive."[34] Iacocca, who had similar qualms about airbags, has moved on: Today, the man who called the Clean Air Act "a threat to the entire American economy" runs EV Global and makes electric bikes. He proclaims the EV's eventual public acceptance to be inevitable.

The prospect of car factories actually closing down because of air-pollution innovations was raised by Earnest Starkman, a GM vice president, in 1972 congressional testimony. If forced to introduce catalytic converters on 1975 models, he said, "It is conceivable that complete stoppage of the entire production could occur, with the obvious tremendous loss to the company, shareholders, employees, suppliers and communities."[35] The cats went on, and the company didn't collapse.

Three years later, GM President E. M. Estes told the Detroit Rotary Club that proposed corporate fuel-economy standards would destroy the "full-sized" automobile. "The largest car the industry will be selling in any volume at all will probably be smaller, lighter, and less powerful than today's compact Chevy Nova, and only a small percentage of all models being produced could be that size," he said.[36] Chrysler concurred. Alan Loofburrow, a vice president of engineering, announced to the U.S. Senate that the Clean Air Act would "outlaw a number of engine lines and car models [by 1979], including most full-size sedans and station wagons."[37]

Big Three executives and their trade associations are on record as saying that acid rain is not a serious environmental problem; that emission standards are impossible to meet; that fuel-economy standards should be rolled back; that the Clean Air Act of 1970 is too stringent; that air bags are unnecessary; that the technology "does not exist" to produce EVs; and many other things that have proven either ridiculous or completely untrue. The amazing thing is that all these assertions were taken seriously at the time, and federal regulation has rarely interfered with auto-industry agendas.

Not that much has changed. A raft of auto executives have been quoted as questioning the science behind global warming, and wondering why their industry should risk billions on unproven assertions. The Global Climate Coalition, which lobbies against action on reducing greenhouse-gas emissions, has had Ford, GM, and Chrysler as members. But the industry is not the united front that it once was. Some companies have broken ranks and supported legislation to reduce carbon dioxide emissions. Others have even called for increased gas taxes. As a general rule, however, auto companies still oppose any interference with what they consider to be their free-market mandate.

THE BILL COMES DUE: PAYING FOR CONVENIENCE

Today, we are as addicted to automobiles as a gambler is to dice, and the love affair has had striking consequences. Public transportation in the U.S. is declining everywhere, with less than 3 percent of Americans using it to get to work. Our private cars, convenient though they may be, are pollution factories. In one year, the average gas-powered car produces five tons of carbon dioxide, which as it slowly builds up in the atmosphere causes global warming. Every gallon of gasoline burned in an automobile engine sends twenty pounds of carbon dioxide, containing five pounds of pure carbon, into the atmosphere. "It's like tossing a five-pound bag of charcoal briquettes out my window every twenty miles or so," writes John Ryan in *Over Our Heads: A Local Look at Global Climate.* He adds that cars and trucks produce by far the biggest share of fossil-fuel emissions (47 percent by one measure). Auto plants are also a significant source of emissions, particularly from their paint shops, though some manufacturers have switched to cleaner water-based paints. According to 1996 EPA data, a single Mitsubishi plant in Normal, Illinois, produced 21.6 pounds of toxic chemicals per vehicle.

Carbon dioxide levels are now more than 25 percent above what they've been for the last ten thousand years, and most scientists think they will continue to rise unless we finally address our tailpipe emissions. At our current rate of carbon dioxide production, scientists predict that global average temperatures will rise two to six degrees Fahrenheit, with an attendant dramatic increase in sea levels, by 2100.[38]

The auto industry is skilled at deflecting blame for all of this. The American Automobile Manufacturers Association (which has now been merged into the international Alliance of Automobile Manufacturers) claimed that today's automobiles are 90 to 96 percent less polluting than their counterparts thirty-five years ago, and that "a 1996 model car can be driven about 60 miles and still give off less smog-forming emissions than a 1965 model parked in the driveway all day with its engine off."[39] But automobile tailpipes still produce a quarter of the carbon dioxide generated annually in the United States.

Obviously, the developed world hasn't done enough to reduce pollution levels overall. Despite the global accord reached on reducing hydrocarbon emissions at the 1992 Earth Summit in Brazil, only Great Britain and Germany have come even close to meeting their 2000 targets. The United States is likely to fall short of its goal by 15 to 20 percent. How could this laudable effort at international cooperation succeed, when cars are the major nonstationary culprits and almost every country is filling its roads with more and more of them?

The international agreement on global warming signed by 150 countries in Kyoto, Japan, late in 1997 makes the effort to cut down on automobile exhaust even more urgent. As the world's largest producer of carbon dioxide emissions, the United States was asked to reduce the greenhouse gases it emits to 7 percent below 1990 levels by 2012. (The Clinton administration had sought a cut to, and not below, the 1990 baseline.) To show where industry priorities lay, *Autoweek* labeled U.S. Senator Chuck Hagel (R-Nebraska) "a friend of the car business" because of his vociferous opposition to Senate ratification of the Kyoto accords.[40]

Meeting the Kyoto goals will not be easy, and the auto industry is in no way ready to do its part. A start would be a tightening of what's known as corporate average fuel economy (CAFE), the federal standard for cars and trucks. The average twelve-mile-per-gallon gas guzzler emits four times as much carbon dioxide as does a fifty-mile-per-gallon compact. The auto industry has fought tenaciously against any attempt to tighten CAFE and, at least until 1998, when California cracked down, succeeded in shielding its beloved sport-utility vehicles from any tough regulations. The result is that actual fuel economy has declined since 1988, as the Big Three have switched from producing fuel-efficient small cars to more profitable—but gas-guzzling—light trucks and sport-utilities.

The Coalition for Vehicle Choice (CVC), an auto-industry lobbying group sponsored by the carmakers, campaigns aggressively to repeal the CAFE standards. The CVC claims that CAFE has, among other things, caused 2,000 deaths and 20,000 injuries every year by forcing people into smaller cars than they would otherwise drive. CVC president Diane K. Steed (who actually administered the standards as head of the National Highway Traffic Safety Administration during the Reagan years) said in 1995 congressional testimony that CAFE "is a program which delivers little if any benefits to consumers, but which exacts a high cost in terms of dollars, lives, and choice."[41]

In public forums, auto executives have either derided the science behind global warming or simply claimed there wasn't enough information to make a judgment. Behind the scenes, they've lent their clout to public relations campaigns dismissing the whole matter. But now that's beginning to change. Toyota was the first auto company to announce, in the spring of

1998, that it was joining with other heavyweights such as British Petroleum, Enron, United Technologies, and Lockheed Martin in an alliance to fight global warming, not the scientists who are warning the world about it. Toyota's support, if not its cash, will go to the Pew Center on Global Climate Change, funded with a $5 million start-up grant from the Pew Charitable Trusts.

GM, which partners with Toyota on alternative-fuel research, moved toward acknowledging global warming later in the same year at a joint press conference with the World Resources Institute (though it got hot under the collar when *The Detroit News* headlined its story on the news conference "Global Warming Is Real, GM Says").[42]

Toyota, which has aggressively sought an environmental identification, is just shielding itself against the forces that could batter the auto industry in the twenty-first century. Global warming is one. And, strange as it may seem in an era of exceptionally cheap gasoline, the end of the fossil-fuel era is another.

TO THE LAST DROP: ARE OIL SUPPLIES RUNNING LOW?

If the auto industry usually gets what it wants, and is only minimally affected by regulation and environmental concerns, why the sudden interest in fuel-cell engines and efficient hybrids? While there are many factors, one not insignificant one is the worry that fossil-fuel supplies, always finite, are finally running out. The same concern is evident in oil-company planning, which includes diversification and consolidation to cushion the shock of a supply crunch. Oil scarcity is an alien concept to motorists who've become used to paying less for gas than for bottled water. In real dollars, we're paying only half as much for gasoline now as we were in 1980. But as *Scientific American* observed in a recent special report, "the end of cheap oil" may be at hand. Demand, the magazine says, could soon start to exceed supply, a problem exacerbated by the concentration of most remaining large reserves in a few Middle Eastern countries. What's more, some experts say, the size of many countries' oil reserves has been systematically exaggerated for political and economic reasons. (A competing theory, advanced in *The Wall Street Journal,* has it that known oil reserves remain relatively stable because oil is actually a renewable resource, continuously produced by the extremely hot temperatures and high pressures under the surface of the earth.)[43]

We've grown accustomed to a steady, seemingly inexhaustible supply of inexpensive petroleum, but the history of oil drilling—indeed, the history of *any* use for what was once called "rock oil"—is surprisingly brief. As late as the 1850s, people scratched out a bare existence by soaking up surface oil from springs around Oil Creek in Pennsylvania. The few barrels they produced were used to make patent medicines. An enterprising Yale chemistry professor named Benjamin Silliman Jr. was the first person to figure out that oil supplies could be reached by drilling (as was then done for salt), and he hit his first strike in 1859.[44] Only a little more than one hundred years after "the light of the age" was first proclaimed, we've used up half the world's known oil supply.

"The question is not whether, but when, world crude productivity will start to decline, ushering in the permanent oil shock era," says oil analyst L. F. Ivanhoe.[45] The World Resources Institute (WRI) predicted in a 1996 study entitled "Oil as a Finite Resource" that world production could peak as early as 2007, at the low end, or 2014, at the high end. This is not just wishful thinking by environmentalists. John F. Bookout, a former president and CEO of Shell Oil whose comments are cited in the WRI report, is in basic agreement with Ivanhoe and

WRI, projecting that production will peak at 75 million barrels a day "around the year 2010."

Energy Investor magazine is even more pessimistic. "The next oil crunch will not be temporary," Colin J. Campbell writes. "Our analysis of the discovery and production of oil fields around the world suggests that within the next decade, the supply of conventional oil will be unable to keep up with demand."[46]

Though some environmentalists, including car industry critics like Ralph Nader and Amory Lovins of the Rocky Mountain Institute, aren't convinced that the end of Big Oil is nigh, the usually optimistic environmental writer Gregg Easterbrook has been doing the math and thinks the wells will indeed come up dry soon. The world has used up 800 billion barrels of oil, he writes, with between 1,000 and 1,600 billion barrels remaining in production-feasible reserves. At the current rate of consumption, it would take sixty years to use up the remaining supply. However, Easterbrook says, "world consumption is not standing still, but increasing. Current global petroleum use, already a mind-bending 71 billion barrels a day, is rising at almost 2 percent a year."[47]

Oil consumption in the Third World is skyrocketing. In Taiwan, for instance, oil imports in 1997 were up 70 percent. In China, they rose 37.5 percent in 1996. "We're either going to have to find huge new deposits soon—which is essentially impossible—or we're going to see sharply rising prices, shortages and economic disruption," writes oil analyst Richard Reese.

Although, as WRI points out, there is a "general failure to appreciate how little time remains before global [oil] production begins to decline," the auto industry is well aware of such dire predictions. It could hardly have missed the near hysteria that developed when, in early 1999, gas prices started to rise sharply in southern California and elsewhere. A desire to hedge their bets is one reason the car companies are pouring millions into EVs and, especially, fuel cells.

THE END OF THE GASOLINE ERA?

The gas-powered automobile has become a fixture of modern life, and it won't disappear easily. "Deposing the car from its dominion over the earth is a radical, even revolutionary move," admits Jane Holtz Kay in *Asphalt Nation.* We'll certainly still be driving cars in 2025 and beyond. But just as the EV temporarily faded into history in the 1920s, the internal-combustion car, befouling the air through its exhaust pipe, may soon begin its slow fade to black.

The end won't come through any one single factor. We won't suddenly run out of oil and turn the pumps off, for instance. Carmakers won't drastically curtail emissions in a genuflection to the gospel of global warming. Fuel-cell automobiles won't emerge fully formed and instantly corner the car market. Federal legislators won't ban the tailpipe. But there are still many pressure points hastening a landmark shift away from internal combustion.

EVs won't completely solve our fossil-fuel problems. Early fuel-cell cars may well run on them. Hybrid cars will mostly burn diesel, though they'll do it very efficiently. And as critics point out, battery EVs are reliant on electricity from utility-owned power plants that often burn oil or coal. But the fact is that, as the Union of Concerned Scientists reports, it would take thirteen EVs getting their power from a fossil-fuel-burning grid to equal the volatile organic production of the average gas car, six to equal its nitrogen oxide production, and 1,440 to match its carbon monoxide.[48] Another benefit is that electric motors are as much as 90 percent efficient, compared to the 25 percent efficiency of the average gas engine.[49] And utilities like

Southern California Edison claim they could run millions of EVs on their unused off-peak electricity, without building any new plants.

There will be teething problems in any switch to EVs. But what's got the automotive fraternity interested is a vision, almost within technical reach, of an automobile that's the industry equivalent of a smokeless cigarette. If hydrogen could be produced from renewable resources, like photovoltaics and wind power, cars would become virtually emission-free, and greenhouse gas production from vehicles would drop as much as 90 percent.[50]

The notion that internal combustion is in its last days is not solely an environmentalist's wishful thinking, but also the considered opinion of a growing number of top-level auto executives, who predict that outcome in general terms, but are less willing to put a date on it. The invention of the automobile did not immediately retire the horse, and, similarly, the first clean cars will seem like exotic novelties. We're not likely to see a Pied Piper leading 200 million gas burners to the junkyard.

I've been reporting on the auto industry for fifteen years, and in that time have experienced considerable dismay at what I perceive as a willful denial of serious environmental concerns. In micro form, it's revealed in snide asides at press conferences; on the macro level, it's seen in industry lobbying and slick public relations campaigns. But now, for the first time, I see evidence of changing thinking. Environmentalists are getting a hearing in Detroit, and global warming is finally being taken seriously. This doesn't mean that General Motors will get any friendlier to the notion of an environmentalist in the White House, or that California's regulators can let industry lead the way. The change is glacial, and incremental, but it is undeniably there.

CHAPTER 3

ENGINES OF INGENUITY: NEW TECHNOLOGIES FOR THE CLEAN CAR

AFTER INVESTIGATING THE LONG, colorful, and occasionally sordid history of the automobile's interaction with the environment, I began to sense that we were once again at a crossroads, looking into an uncertain future. Like the Mr. Jones Bob Dylan sings about in "Ballad of a Thin Man," I knew something was happening, but I wasn't sure exactly what it was. Instead of trying to interest me in the sweep of a redesigned fender, car people were suddenly talking excitedly about a host of exotic technologies, some new and others just resurrected. It was to some extent bewildering, because for my entire lifetime the industry had focused relentlessly on internal combustion, making only minor changes from year to year. Some engines, such as Chrysler's famous Slant Six, stayed in production for almost three decades. I decided I needed to know more about how these new power plants worked.

I also began to see that the gasoline engine carried considerable baggage with it, including trillions of dollars of entrenched oil-related infrastructure. Multiply the corner gas station and Express Lube outlet by the tens of thousands, and you have a huge investment. Could that weighty assemblage be moved? I found that history was instructive, up to a point. Go back a hundred years, and you arrive at a time when gasoline was by no means the fuel of choice, and alternative energy sources competed fiercely. By learning from the past, could we go back to the future?

In 1900, when there were only a few thousand motor vehicles registered in the United States, the public could choose from steam, electric, or gasoline automobiles, all of which had passionate partisans, industrial backers, and embryonic infrastructures. There was as yet no clear winner. Although we take it for granted today, a gasoline-based transportation system was by no means a foregone conclusion then. To a public used to horses, the idea of sitting on top of a highly flammable gasoline tank, making forward progress by a series of dramatic explosions, was not instantly appealing.

Since the knowledge that car exhaust could be dangerous was at least fifty years in the future, pollution was not at issue. Indeed, none of the early advertisements for electric cars touted their clean air benefits, though the fact that they ran silently and didn't soil their operators' clothes certainly got some attention from the copywriters.

A century later, the auto industry is at a similar crossroads, with competing technologies vying for a place on twenty-first-century roads. Just as in 1900, there will be winners and losers and technical breakthroughs that lead to hairpin turns in corporate thinking. It's impossible to say which new propulsion system will emerge triumphant, but it *is* becoming clear that the internal-combustion engine will become a historical artifact, after what may turn out to be, in history's long sweep, a relatively brief hundred-year run. What's more, the new, clean cars will emerge not from a tinkerer's garage, but from the well-funded research labs of the same big auto companies that initially fought their introduction.

The new technologies range from the relatively familiar (battery EVs) to the completely exotic (fuel-cell EVs), though all are, in essence, electric cars. Each presents unique challenges and opportunities on the road to a successful commercial introduction. The clean cars will be lauded as high-tech wonders, and they'll deserve the praise, but they are firmly rooted in the nineteenth century, when the basic research that governs their drive systems was conducted. Modern materials, electronics, and streamlined manufacturing processes make physically possible today what was only theoretical then.

THE FUEL-CELL EV

The fuel cell, first demonstrated in principle in 1839, is an excellent example of a scientifically proven technology that could find no practical application during the inventor's lifetime. Indeed, the inventor made it plain that actual uses for his "gas battery" didn't interest him in the least.

Sir William Robert Grove (1811–1896) was a larger-than-life figure of the type that proliferated in nineteenth-century England. Educated at Oxford and trained as a barrister, Grove became famous or, perhaps, infamous as the unsuccessful defender of William Palmer, the notorious "Rugeley poisoner," who used strychnine on more than a dozen victims. Grove was later to become a judge. But he was also scientifically oriented and would sometimes get so sidetracked in patent cases that he'd end up suggesting improvements in the products' designs, rather than restricting himself to mere legalisms.

When he wasn't presiding in court, Grove could be found in his laboratory, where he made several important improvements to the design of storage batteries. He also appears to have been, in 1845, the first scientist to describe and build a filament light. Grove, who was knighted in 1872, could have rested his scientific reputation on these laurels, but he also invented the fuel cell, describing in some detail how the chemical combination of hydrogen and oxygen could be used to produce electricity.

In 1842, Grove lectured on the gas battery's properties in London. His science was sound, based on the idea that it should be possible to reverse the already well-known electrolysis process and get electricity out rather than put electricity in. In electrolysis, which is widely used in metal plating, a current is introduced into an electricity-conducting liquid known as an electrolyte, where it flows between two electrodes and causes chemical changes. Grove proved that his reverse principle worked, generating a powerful current in his laboratory, but the practical applications of his invention failed to stir him. "For my part," he told the Chemical Society in 1891, "I must say that science to me generally ceases to be interesting as it becomes useful."[1]

How a Fuel Cell Works

Fuel-cell technology can be compared to that of a car battery, in that hydrogen and oxygen are combined to produce electricity. But batteries store both their fuel and their oxidizer internally, meaning they have to be periodically recharged. Like a car engine, the fuel cell can run continuously, because its fuel and oxygen aren't sealed up inside it. Pure hydrogen gas—or hydrogen extracted from a fuel like methanol or gasoline—is fed to the anode, one of two electrodes in each cell. The process strips the hydrogen atoms of their electrons, turning them into positively charged hydrogen ions, which then pass through an electrolyte (depending on the type of fuel cell, it can be phosphoric acid, molten carbonate, or another substance) to the second electrode, known as the cathode.

Meanwhile, the negatively charged electrons, which can't travel through the electrolyte, move to the cathode via wire.

This movement produces electric current, the intensity of which is determined by the size of the electrodes. At the cathode, the electrons are brought back together with their ions and combined with oxygen to produce the fuel cell's major by-product, water. The other by-product is waste heat, which in some applications can be captured and reused in a process called cogeneration.

A cell generates just under one volt, but fuel cells can easily be grouped in "stacks" to produce more voltage. In a pure hydrogen fuel cell, emissions are nonexistent, but some release of pollutants is inevitable in a car that "reforms," or extracts, its hydrogen from a fossil fuel. Even so, levels are quite dramatically lower than what comes out of the world's dirty tailpipes. Unlike the modern car engine, with its noise, heat, and rapidly spinning parts, the fuel cell is just an enclosed box, with no moving parts and no noise. It's not much to look at, but its implications are vast, not only for transportation but also for the entire energy constellation, since fuel cells work even better in stationary applications—such as home power generation—than they do in cars.

Although there are several different kinds of fuel cells, only one type, the proton-exchange-membrane (PEM) cell, is seriously being considered for cars. The PEM cell, which was developed for the Gemini space program by General Electric in the early 1960s, has no equal in terms of size, low operating temperatures, adjustable power outputs, or quick starting. Breakthroughs at the federal government's Los Alamos National Laboratories in the 1980s made the practical PEM cell possible, by drastically reducing by up to 90 percent the amount of precious metal catalyst needed to coat the cell's ultrathin polymer membrane.

Although the modern work on fuel cells is all pretty recent, the technical problems of the cells themselves have been mostly worked out, and Ballard Power Systems, the British

Columbia–based leader in this technology, intends to have a car-sized power unit ready to go, at prices comparable to internal-combustion engines, before 2002. That by no means puts a fuel-cell car on the showroom floor. The real obstacle, the hurdle keeping engineers in both industry and government up nights, is the fuel itself. Will fuel cells run on pure hydrogen, meaning they'll have to carry a high-compression tank of this highly flammable gas on board, or will they require (at least as an interim step) a reformer to extract hydrogen from a fossil fuel, probably methanol? Although most environmentalists favor the "direct hydrogen" approach because it's cleaner, the auto industry seems bent on retaining its familiar liquid fuels, and the first fuel-cell cars will probably run on them.

In late 1997, a joint project of Boston-based Arthur D. Little Inc.; Latham, New York's Plug Power; and the U.S. Department of Energy publicly demonstrated a gasoline reformer, a stunning achievement, since gasoline is among the hardest fuels to reform. Gasoline contains sulfur, which poisons fuel cells, but Epyx, the Arthur D. Little spin-off company that's working on the gas reformer, captured the sulfur before it got to the cell, using a device similar in principle to a catalytic converter.

Bob Derby is the marketing director of Epyx. "We envision a reformer that can work with multiple fuels and can be changed on the fly to use gasoline, ethanol, or methanol," he said. "You could compare the unit to a portable generator, except its efficiency levels will be much higher and its emissions levels much lower."

Environmentalists aren't exactly jumping up and down celebrating the Epyx achievement, since it potentially postpones the inevitable day of reckoning with fossil fuels. "The environmentalists didn't like it," says Jeffrey Bentley, the Epyx CEO who invented the reformer. "But I came to the conclusion that you've got to adapt to the current infrastructure."[2] Epyx was dealt a blow in early 1999 when Chrysler, which had been the strongest proponent of gasoline reforming, changed its mind and switched to methanol. The move reflects Chrysler's merger with methanol-boosting Daimler-Benz, but it's also an indication of the great complexity involved in reforming a highly refined fuel like gasoline.

A reformer adds more weight to a car that must be as light as possible, and it's a complicated, miniature chemical factory. What's more, "reformed" hydrogen is not pure, and isn't likely to deliver the same performance as hydrogen gas. It may come down to a question of infrastructure. If fuel-cell cars run on gasoline, obviously we don't have to change the local gas station. But to turn the trickle of hydrogen we produce now for industrial use into a mighty national network could cost hundreds of billions. One scenario is that methanol fuel-cell cars will bridge the gap in the decade or more it might take to build that network. Another, more environmentally friendly, possibility is that gas stations will install miniature electrolysis factories next to their pumps and produce hydrogen from water. Robert W. Shaw Jr., whose Arete Corporation was an early investor in fuel-cell technologies, thinks we may even see photovoltaic cells on the roofs of service stations, cleanly producing power to make hydrogen locally.

The Safety of Hydrogen Fuel

An obvious advantage of gasoline-reformed fuel cells is that they avoid the hydrogen safety question. And that question inevitably arises, in part because of the spectacular 1937 fire that killed thirty-six people and destroyed the German zeppelin *Hindenburg.* The blaze, which occurred as the 240-ton airship was docking in Lakehurst, New Jersey, put an immediate end to zeppelin travel and saddled hydrogen with a nasty reputation it has yet to fully shake.

The *Hindenburg* was not hydrogen-fueled; the extrabuoyant gas, used because helium was not available to the increasingly bellicose Nazi regime, filled sixteen cells in the airship's body and gave it lift. Was the hydrogen on board the *Hindenburg* responsible for the fire? Conventional history has made that case, but retired NASA engineer Addison Bain, a hydrogen specialist, thinks otherwise. After several years of research that included tracking down surviving pieces of the *Hindenburg*'s cotton skin, Bain is convinced that the onboard hydrogen certainly fueled the fire, but it played no role in starting it. The culprit, he believes, was the highly flammable cellulose doping compound used to coat the fabric covering and make it taut. On top of that, he says, the airship's builders used aluminum powder—an ingredient in rocket fuel—to reflect sunlight so the gas in the cells would not expand. Lacquer, another flammable substance, was used to coat the *Hindenburg*'s support structure.

Bain believes that an electrical discharge ignited the zeppelin's skin after it docked, and that the heat from the fire then burst the hydrogen cells and ignited the escaping gas. (The gas leak caused the still-buoyant nose of the airship to rise, as it is seen to do in many famous photos of the disaster.) "I guess the moral of the story is, don't paint your airship with rocket fuel," says Bain.[3]

The safety of hydrogen gas carried on board a modern motor vehicle is, of course, an entirely different matter than speculating about what caused a long-ago tragedy. There are some, even in companies that make fuel cells, who speculate that hydrogen is simply too volatile, too dangerous, to ever be safely domesticated for cars. Peter Voyentzie of Danbury, Connecticut's Energy Research Corporation, which makes large stationary-fuel-cell power plants, is skeptical about automotive applications. "Hydrogen is a strange beast," he says. "It's the smallest molecule, and it leaks out of everything. You also can't see it burn. In a car, it has to remain stable through collisions and constant agitation. That's a lot to expect."

But those worried about hydrogen's propensity to burn might want to consider that fifteen thousand cars are destroyed by engine fires every year, and five hundred people die from auto accident–related burns. Gasoline is itself highly volatile, a fact that clinched a considerable number of EV sales in the early days of motoring. Today, we're so familiar with gasoline that it no longer seems very dangerous (even as we watch Hollywood stunt cars explode on-screen).

Hydrogen is, indeed, a strange beast. When spilled, it simply escapes upward instead of puddling like gasoline and presenting an ignition hazard. It is odorless, its flame is invisible, and it emits very little radiant heat, so people standing next to a hydrogen fire might not even be aware it's there. Even in diluted form, hydrogen will burn easily, but unless you're in physical contact with the fire, it won't hurt you. Another safety advantage of fuel-cell cars is that they don't burn their fuel, making ignition less likely in the event of a collision or leak. But a spark generated by the friction of a crash could set it off.

"Hydrogen is a lot safer to carry around than gasoline," says Amory Lovins. "If we had a hydrogen economy and someone proposed introducing gasoline, it would be prohibited as way too dangerous. I would for a darn sight rather be in a crash in a hydrogen car than in a gasoline car, from the fire and explosion perspective. Hydrogen is fifty-two times more buoyant, and thirteen times more diffusive, than gasoline. Victims generally survive better in a hydrogen fire, because they're not burned unless they're in it."

Hydrogen tanks for cars have to be sturdy indeed, stronger than the car itself, and the best of them are made of reinforced composites with an aluminum liner. Tank makers have even gone so far as to drop their products from fifty feet in the air, and they've remained intact, without leaks. Further strength may be necessary if internal tank pressures, now normally 3,000 pounds per square inch, are nearly doubled to 5,000 pounds. One problem, which will probably have to be addressed with venting and gas-detecting sensors, is the potential for hydrogen to accumulate in the roof of garages and in tunnels.

Hydrogen's safety problems shouldn't be minimized, but they shouldn't disqualify the fuel from consideration. Like gasoline, hydrogen can be dangerous. And, also like gasoline, we can learn to use it as safely as possible.

EVS WITH BATTERIES

As clean as they are, fuel-cell cars with reformers can't claim to be zero-emission vehicles, since they don't meet the standards set by California for the environmental paragons it wants to see on state roads. For the most part, ZEVs are battery-powered electrics that rely on the utility grid. As we've seen, EVs have a proud history, one that can be traced to an inventive Vermont blacksmith named Thomas Davenport, whose revolutionary rotary electric motor powered a carriage, laden with eighty pounds of ballast, in the late 1830s.[4] Davenport got some favorable newspaper notices, but like his contemporary steam pioneers, he was ridiculed by his horse-dependent neighbors and made no money from his country fair novelty.

Before there could be successful EVs, there had to be reliable electric storage, and that takes us back even further than Davenport, to the work of the Italian count Alessandro Volta, whose voltaic pile of 1800 was the world's first battery. Volta, a large and active man who, it was said, "understood a lot about the electricity of women," developed his cell using round plates of copper and zinc as electrodes, with cardboard soaked in salt water as his electrolyte. Current flowed, and the world cheered. Volta was even called to France by Emperor Napoleon in 1801 to demonstrate his invention. Volta's battery produced dependable electricity, a great boon to scientists, but this "primary" cell couldn't store its power for any length of time and so was useless for transportation.

EV enthusiasts really owe their thanks to the Frenchman Gaston Planté, who in 1859 developed the first lead-acid "storage" battery, capable of extended use and repeated recharging. When one connected an electric current, the chemical conditions within a depleted cell were reversed, and the battery was ready for duty again. Descendants of Planté's "secondary" battery are installed in every car on the road today. And when multiple storage batteries are connected to an electric motor, the result is an EV that can travel, practically noiselessly, as long as its charge lasts. It's not surprising that the first electric carriages appeared almost immediately after Planté's batteries were commercialized.

The battery-powered EV is old technology. By 1905, they were a common sight on the public roads, and they were reasonably reliable. More than ninety years later, most people would expect them to have kept pace with the times, and be several orders of magnitude better than they were then. Unfortunately, they aren't. In the early 1980s, *Machine Design* magazine tested a 1915 Detroit Electric against the best of the EVs then on the market, a motley assemblage of kit cars and conversions. The quality-built Detroit, which offered a drawing-room ambiance complete with overstuffed armchairs for five passengers and curtains, could cruise for eighty miles (albeit at a top speed of twenty-five miles per hour). The modern cars couldn't do much better, so the editors deemed the Detroit a "best buy."[5]

The internal-combustion engine has benefited from more than a century of continuous development and is, even from the point of view of a committed EV engineer, a very evolved technology. The modern gas car is a very effective, if very polluting, tool, capable of holding sustained speeds in reasonable comfort and safety, with a range of 350 miles or more. By contrast, until GM's EV1, no major corporation had put its full research and development muscle into EVs since the last straggler manufacturers died out in the early 1930s. Big auto companies have, at least until recently, relegated EV work to a background activity, and backyard tinkerers and small start-ups have lacked the capital to produce fully realized challengers.

While today's EVs have benefited enormously from modern technology, particularly advances in electronics, their single biggest problem remains a limited range. When equipped with an affordable variation on the lead-acid batteries Gaston Planté invented in 1859, the otherwise ultrasophisticated EV1, introduced by General Motors with much fanfare in 1996, can go no farther than seventy miles in city driving before needing a recharge. Environmentalists love battery EVs, which have such a friendly image that people call them "solar cars" in the mistaken belief that most of their energy comes directly from the sun's rays. But the battery car, while environmentally wonderful, has severe handicaps in the race to become the world's mainstream automobile.

It's the batteries that are the problem. Range-limited lead-acid packs—the most affordable option—were until recently the standard for most EVs, though the more exotic nickel metal hydride (NiMH) packs, pioneered by a self-taught engineer named Stanford Ovshinsky, have rapidly taken their place.

The common 12-volt lead-acid battery is made up of six identical cells, each containing two electrodes and positive and negative lead plates bathing in an electrolyte solution of sulfuric acid and water. This old friend has a number of advantages: It's proven technology (a million are sold every year just for golf carts), it's not expensive to manufacture, and it's long-lasting. But the energy density of lead-acid batteries, the amount of power they can deliver on a charge, is poor when compared to NiMH and other contenders.

The United States Advanced Battery Consortium (USABC), a Department of Energy program launched in 1991, has been described as "a Manhattan Project for advanced batteries."[6] The idea is to consolidate DOE technology grants in the hope of producing a "super" battery breakthrough and get viable EVs on the road sooner. Since 1992, USABC has invested more than $90 million in the nickel metal hydride batteries Ovshinsky first demonstrated, using a nickel electrode scrounged from a Kmart battery, in his laboratory in 1982.

Ovshinsky's Ovonic batteries are much cheaper to make than earlier nickel battery types, and have an energy density approximately double that of lead-acid. Mark Verbrugge, GM's representative to USABC, says that NiMH batteries can accept three times as many charge cycles as lead-acid, and work better in cold weather—an important factor for EV acceptance in the northern United States.

Are NiMH batteries a "killer ap"? They've certainly proven effective in laptop computers, cellular phones, and video cameras. But they are still quite far from meeting USABC's cost and range targets. The consortium is trying to find a battery pack that can give an EV a reliable range of two hundred miles; Ovonic batteries can take an EV1 just over one hundred miles, and are still several times more expensive than lead-acid.

But, objections aside, NiMH batteries are getting an acid test. In 1994, Ovshinsky's company, tiny Michigan-based Energy Conversion Devices, teamed up with GM to "jointly pursue commercialization" of Ovonic batteries, with former GM chairman Robert Stempel acting as matchmaker. Stempel was ousted from GM in 1992 after the company lost more than $4 billion in 1991. He saw green soon after, first as a consultant to the company he'd just left, then as a "senior technical adviser" to Energy Conversion.

A partnership named GM-Ovonic was set up, and, in late 1998, Ovonic batteries were installed in GM's EV1 and S-10 electric pickup truck, doubling the range of each. Around the same time, Chrysler announced that it, too, would put NiMH batteries, made by GM-Ovonic competitor SAFT of France, in its Electric Powered Interurban Commuter (EPIC) vans, adding thirty miles to their range. Battery cars were definitely getting better, now that some engineering effort was going into them. But they were still having a hard time shaking the reputation, at least partly deserved, that even with a full charge "they're like a gasoline car with the fuel gauge reading empty."

The technology graveyards are littered with dead battery technologies, including sodium-sulfur, a fiasco in early Ford EVs, and zinc-air. Zinc appeared in GM's failed Electrovette EV in the late 1970s and turned up again a decade later in the zinc-air batteries touted by a number of companies, including Israel's Electric Fuel, Ltd. Zinc is cheap, and these batteries have six times the energy density of lead-acid. A car with zinc-air batteries promised to deliver a four-hundred-mile range. Unfortunately, as testing by the German postal service demonstrated, these batteries can't be conventionally recharged. No one has yet successfully described an infrastructure that includes removing full battery packs after each long run on the interstate.

Other battery types are more promising, including lithium-ion, which is showing up in a wide variety of consumer products. But will lithium batteries, which could offer high energy density, long cycle life, and the ability to work in many different temperatures, ever be practical in cars? Unfortunately, like the sodium-sulfur batteries in the Ford Ecostar, lithium-ion presents a fire hazard because lithium itself is highly reactive.

Plastic lithium batteries could prove to be very versatile. New Jersey–based phone giant Bellcore is working to make a practical lithium battery that would be as thin and bendable as a credit card—with its first targets laptop computers and cell phones. "It can take any form or shape," says Christina Lampe-Onnerud, Bellcore's director of energy storage. "Each cell is only a millimeter thick, and you can treat it like fabric—just cut off as much as you need." The plastic batteries are very lightweight, and have undergone some preliminary testing in automotive applications.

In a related development, the Canadian utility Hydro-Quebec, working with 3M, has conjured up a lithium-polymer prototype that it says is the first "solid-state" EV car battery. Like the Bellcore product, this dry battery uses a sheet of polymer plastic in place of a liquid electrolyte. Also pursuing the technology is a team working in the labs at Johns Hopkins University in Baltimore. The Baltimore plastic battery, *Popular Science*'s 1997 "invention of the year," is like Bellcore's in that it can be formed into thin, bendable sheets. These batteries also contain no dangerous heavy metals and are easily recycled. But, as with many new EV technologies, getting the batteries from lab success to production reality is still a long road.[7]

HYBRID GAS-ELECTRIC CARS

If battery EVs aren't going to fly, or at least not fly far enough, why not a compromise? The solution, as more and more car industry executives have seen it, is to use the hybrid gas-electric car as a cleaner and greener interim step, plugging a gap of several years as fuel-cell cars continue development. The approach is complicated by the fact that some hybrids use small direct-injection diesel engines, which present pollution problems of their own.

Like EVs, hybrids are not new, since engineers have long recognized that electric motors are great for accelerating and that gas engines work best at a steady speed. Dual power, which can dramatically increase operating efficiency by exploiting the best performance range of each system, was seen in some of the earliest cars and trucks, though it rarely succeeded commercially.

As mentioned in chapter 1, Dr. Ferdinand Porsche's second car was a fully functional turn-of-the-century hybrid. General Electric also built a hybrid prototype, in 1900, and Germany's Siemens built hybrids until 1910. A fascinating twist on the concept was the Woods Interurban of 1905, which for most uses was a battery-powered electric. Long-distance drivers had the option, however, of swapping the electric power unit (supposedly a fifteen-minute job) for a two-cylinder gasoline engine and drivetrain. The Interurban didn't find many customers. More successful was a Commercial hybrid truck, built in Philadelphia between 1910 and 1918, which used a four-cylinder gas engine to power a generator, eliminating the need for both transmission and battery pack.

Of more recent vintage was the experimental fifty-mile-per-gallon GM 512 of 1969, a very lightweight two-passenger hybrid car whose entire front section flipped open, eliminating the need for doors. Like many modern hybrid designs, the 512 automatically switched from electric power to two-cylinder gas power. Until it reached ten miles per hour, it ran on batteries; from ten to thirteen miles per hour both units operated together. Above thirteen miles per hour, with the hard acceleration presumably behind it, the 512 ran on gasoline.[8] The 512 could only reach forty miles per hour, so it was more of a glorified golf cart than a highway-ready car.

Briggs & Stratton of lawn-mower fame got a lot of press for a unique six-wheeled hybrid prototype it built in 1980 to demonstrate the versatility of its eighteen-horsepower gasoline engines. The battery pack and electric motors powered the rearmost wheels, while the gas drivetrain motivated the middle set. No one seriously considered it for production—it must have been very heavy and would have been quite costly to build.

Lincoln-Mercury never tried to actually sell its Antser, either, though this early 1980s bullet-nosed concept car was quite visionary in anticipating that low weight (twelve hundred pounds) plus a hybrid drivetrain would make a very economical, high-performing, long-range car. The Ford division thought the Antser, whose *Star Trek* doors slid backward on a track, was "the type of vehicle we can expect to find in the late 1980s,"[9] but the prediction was off by at least a decade. Ford updated the Antser in 1996 as an even more dramatically lightweight, low-drag hybrid car called the Synergy 2010, but its wild styling marked it as a show-only enterprise. The first production hybrid, when it finally arrived in 1997, came in the guise of the modest Toyota Prius, not a futuristic show car.

Series and Parallel Hybrids

There are two types of hybrid electric car, series and parallel. In a series hybrid, a small gas or diesel engine is used as a generator, producing power to drive the electric motor and recharge the small battery pack that these cars usually carry. The batteries also store electricity generated through so-called regenerative braking, which is an ingenious method of recapturing up to half the "motion energy" moving cars normally waste (as heat warming up their brake pads) during the decelerate-and-stop cycle. Under "regen" braking, the running gas motor is turned into a recharging generator, an especially useful feature when drivers ride the brakes in stop-and-go city traffic.

Parallel hybrids, the type currently going into production in the U.S., have two discrete power systems: Both the gas and electric motors can drive the car, and when they work together they provide the kind of power needed for fast American driving. In one type of parallel hybrid, the car starts out in battery-powered mode, then automatically fires up its internal-combustion engine when the batteries lose 40 percent or more of their charge. The parallel hybrid's computer chooses which type of power is "on line," and makes barely detectable switches between the gas and the electric mode, or combines them for quick acceleration. Another parallel type—the Honda Insight is an example—runs mostly on a small gas motor, switching on the electric for added boost. Many hybrids also have a continuously variable version of the automatic transmission, which cuts down on energy lost during shifting.

Flywheel Hybrids

An intriguing variation on these themes is the flywheel hybrid, as seen in the TurboFlywheel Saturn that California-based Rosen Motors publicly tested in 1997. "This is a milestone in automotive history,"[10] proclaimed chairman Benjamin Rosen, whose day job is as chairman of Compaq Computers. "To us, this demonstration is akin to the first flight of a jet aircraft," said his brother, Harold, the engineer in the family and a former Hughes Communications vice president. The Rosen car stored electrical energy in its flywheel, a furiously rotating 60,000-rpm carbon fiber cylinder operating in a vacuum. Unlike a chemical battery, the flywheel, an old concept based on the potter's wheel made high tech by space-age materials, stores kinetic energy: After being charged up by the car's gas-powered turbine, it can keep turning for weeks, releasing its pent-up power to assist acceleration.

There was a lot of interest in flywheels for a time. Film star Kevin Costner and his brother, Dan, have been major investors in U.S. Flywheel Systems, a company headed by a former NASA scientist. U.S. Flywheel's prototypes spin at 100,000 rpm. A fleck of dust on the flywheel's rim would travel at 3,700 miles an hour, the speed of a bullet.[11] The extremely high speeds raised safety questions (despite Rosen Motors' "containment system" and "high-strength straps"). What would happen to that 100,000-rpm, fifty-pound flywheel if, after an accident, it suddenly became a projectile?

Rosen Motors confidently predicted that it would have a flywheel car capable of zooming to sixty miles per hour in six seconds ready for the production line by 1998, but instead the company shut down in late 1997, having gone through $24 million, mostly of Ben Rosen's money, and failing to attract a major automaker partner.

Hybrids Reach Production

Hybrid car experiments go back one hundred years, but almost all of them were unique prototypes or stillborn marketing efforts. What's so exciting about Honda's Insight and Toyota's Prius is that both are entering the U.S. market backed by the full weight of some very savvy auto giants.

The system in Toyota's breakthrough Prius, which first appeared at the Tokyo Motor Show in 1997 and came to the U.S. in 2000, incorporates elements of both the parallel and the series hybrid. The engine, a fuel-efficient, low-revving (its power peak is at 4,000 rpm) 1.5-liter four-cylinder, delivers its output both to a generator and to the wheels, through an electronic power splitter. The engine shuts down automatically when the car is stopped, which means it won't pollute when stuck in traffic jams. An engine-management system decides how the power is divided, and if it makes the right decisions the Prius (which also has regenerative braking) can achieve its stated sixty-six miles per gallon fuel economy.

Basically, when the car is starting out or moving slowly, the engine is turned off and the batteries are in use. At normal running speeds, the gas motor is started, sending power to the generator and to the road. When the Prius accelerates abruptly, horsepower from the engine is aided by extra power drawn from the batteries. As the car slows down or brakes, the motor becomes a generator, capturing the kinetic energy of the wheels and recharging the batteries.

In Honda's two-seat hybrid, the Insight, a thin electric motor-generator acts as a booster or supercharger for a small gas engine, which does the work of driving the wheels. If the gasoline engine has a future, it's in very efficient power plants like this one, which combines a very small displacement of only one liter with a very lean-burning direct-injection system that squirts fuel right into the combustion chambers. When the engine is combined with a power-storing ultracapacitor and a thin brushless DC electric motor in a platform Honda calls Integrated Motor Assist, the result is a fairly mind-boggling seventy miles per gallon.

Honda, which marketed and then killed a battery-powered car in California, was obviously trying to steal some of the thunder from archrival Toyota's Prius by getting a hybrid on the market first. The Insight, with a very lightweight 1,740-pound aluminum body, went on limited sale worldwide at the end of 1999. The car, which has stylish fender skirts like GM's EV1, looked something like the sporty, two-seat Honda CRX. The Insight has obvious youth appeal, but it isn't a road rocket. I drove the Insight on a 140-mile round trip through some picturesque Maryland countryside. If the car had been battery powered, I would have had to stop and recharge halfway through. Instead, I barely dented the gas gauge.

The Insight won't win any stoplight drag races. Acceleration is roughly comparable to a stock Civic's, around 12 seconds to 60 mph. Because the gas engine is always engaged (unlike the Prius, which can switch between gas and electric modes), there's no steep learning curve in getting behind the wheel. When the battery motor engages on hills, it's only detectable through a dash-mounted gauge. Much more noticeable is the Insight's gas-saving "idle-stop" ploy of shutting down completely when stopped at traffic lights. The car can restart in a tenth of a second, and does so as soon as the driver's foot hits the clutch.

The Insight is a very small car whose diminutive size was particularly apparent when I pulled up alongside a Chevrolet Suburban at a traffic light. But it's not spartan, and it makes allowances for the long-distance traveler. Will the kids care that their coupe gets more than double the gas mileage of Dad's Oldsmobile and meets California's ultralow emissions standards? Environmentally concerned consumers are more likely buyers.

On a parallel track, Honda is trying to make the gasoline engine socially acceptable. It has produced a Z-LEV version of the 2.3-liter four-cylinder engine found in the Accord that, it claims, is nearly pollution-free, with emissions of carbon monoxide and nitrogen oxide down to 10 percent of California's very tough standards. "In some high smog areas like Los Angeles, the Z-LEV's tailpipe emissions can be cleaner than the surrounding air," says Honda.[12]

The Z-LEV engine will "drive a stake through the heart of the electric vehicle," claimed one EV-unfriendly analyst.[13] However, the demands of making a lab engine like Honda's not only ready for production but able to withstand the rigors of day-to-day highway pounding mean its actual street performance could be much worse. With graphs and charts, the Union of Concerned Scientists shows that "real world" emissions are usually much higher than the standards set by Detroit and Washington.[14]

GM, covering its bets and undoubtedly scrambling to respond to the surprise appearance of the Prius at the 1997 Tokyo Motor Show just the month before, announced two entirely different hybrids at Detroit's North American International Auto Show in early 1998. GM chairman Jack Smith and his team from Advanced Technology Vehicles, all togged out in matching green sweaters, put on quite a show. Smith opened it up by driving the new NiMH battery version of the EV1 through a curtain, then hopping up to the podium and telling the assembled journalists that "electrics are here to stay" and that "issues like global climate warming and energy conservation demand fundamental change from all industries in all nations." He added that "hybrids may be a midterm solution, and fuel cells offer strong potential as the best solution long-term."

At the Detroit show, Ken Baker, the talented manager whose diligence brought the EV1 from concept to production reality (and who has since left GM for Energy Conversion Devices), introduced the company's parallel hybrid. The exotic drivetrain is built into a four-seat, nineteen-inch stretch of the EV1 body that looks a bit like a futuristic Citroën. The all-wheel-drive car, set to be production-ready by 2001, can do zero to sixty in seven seconds, so Baker was fully justified in calling the car "a twenty-first-century hot rod."

The parallel car, which also offers eighty-miles-per-gallon fuel economy and a 550-mile range, uses a 75-horsepower Isuzu-built three-cylinder direct-injection diesel engine, coupled with a 137-horsepower electric motor. The parallel car starts in front-wheel-drive electric mode, but when the driver hits the accelerator the diesel joins in, giving the car four-wheel-drive capability and a potential 219 horsepower. Things get more complicated when you factor in the car's multipurpose 6.5-horsepower electric motor-generator, which starts the diesel engine, provides regenerative braking, adds an extra boost of driving power, and recharges the forty-four NiMH batteries (which never go below 80 percent of full charge).

In the simpler sixty-miles-per-gallon series version, which with less performance and range is unlikely to go into production, the same 137-horsepower electric motor drives the front wheels, and a gas turbine engine derived from aerospace technology works as a generator, keeping the same forty-four NiMH battery array charged. The turbine kicks in when the charge in the battery pack drops below 40 percent, which usually happens after about twenty-five miles on the road. GM estimates the car will go from zero to sixty in nine seconds and cover 350 miles before needing a fill-up. A dash-mounted switch locks out the gas turbine, turning the hybrid into a zero-polluting EV with a range of about forty miles.

It may be that GM produces none of these cars, but instead develops something entirely new, probably arising from the wide-ranging five-year partnership with Toyota that the company announced in 1999. Both fuel-cell and hybrid technology are part of the collaboration. It's no great stretch to imagine a fuel-cell Chevy Blazer or even a hybrid GMC Yukon, developed with substantial engineering input from Toyota, plying America's roads in the not-too-distant future. "I think the car companies want to put the new technologies like fuel cells into the familiar physical model—big vehicles that will fit all the kids in the back seat and still tow a trailer," says Robert Massie, executive director of the Coalition for Environmentally Responsible Economics (CERES). Massie's group convinces corporations to commit to a set of green principles, and GM is one of the signatories.

Chrysler used as many theatrics as GM to show off its hybrid Intrepid ESX2 at the same 1998 Detroit show. Tom Gale, a product-strategy vice president, staged a mock game show, with "contestants" Albert Einstein, Madame Curie, and Leonardo da Vinci to answer the question "What's the Big Deal?" about hybrid cars. GM dramatized its cars of the future by having little kids ask probing environmental questions. Not to be outdone, Chrysler had a five-year-old roll out in a plastic pedal car. The ESX2 also has a plastic body, you see. It's made of six pieces of thermoplastic polyester, the same material found in plastic soda bottles.

In prototype form, the ESX2, a beautiful styling exercise that carries on the sleek "cab forward" design of recent Chrysler products, weighs only 2,250 pounds, more than a thousand pounds less than a production Intrepid. It's also more aerodynamic, though the extreme streamlining, both front and rear, doesn't do much for its headroom.

Chrysler calls the ESX2 a "mybrid," or "mild parallel hybrid," since, like the Honda Insight, the three-cylinder direct-injection turbo-diesel engine runs continuously and does most of the work. The electric motor is there mainly to give the car extra acceleration, move it in reverse, and power the accessories.[15] The ESX2 driver doesn't have to sacrifice creature comforts for the car's estimated seventy miles per gallon: The car carries five people in luxury and holds more luggage than any other Chrysler product. It's futuristic, too. Turn the ignition key, and holographic gauges appear from behind a Plexiglas dashboard. That dash also holds a palm-top computer that can download e-mail or surf the Web. Even the shifter looks like a computer mouse.

As good as the ESX2 is, there are obstacles in its path. Hybrid drives are so complicated that the car carries a $15,000 cost penalty, meaning it would probably be the most expensive mid-sized sedan on the market. And while the diesel engine is very efficient, it is still a significant emitter of particulates and nitrogen oxides, presenting a red flag to some clean-air regulators.

Ford has a hybrid, too, of course, and it's based on a car even lighter than the ESX2, the 2,000-pound aluminum and composite P2000 that's been making the show rounds. In basic "family car" form, with a 1.2-liter direct-injection engine, the P2000 can get sixty-three miles per gallon, and it does much better in the gas-electric hybrid version Ford showed at the end of 1998. The same basic structure houses Ford's experimental fuel-cell cars.

Hybrid cars are hardly a slam-dunk technology for America. U.S. motorists do a lot of highway driving, and at a seventy-miles-per-hour cruising speed the electric motor becomes just more dead weight to carry around. In superhighway driving tests conducted by some very interested American car companies, the Toyota Prius did much worse on fuel economy than the sixty-six miles per gallon it claims in Japan. And fuel economy is hardly a major issue with the driving public anyway, with gasoline so inexpensive.

"In the U.S., with gasoline cheaper than bottled water, where's the incentive for the Prius?" asks Robert Purcell Jr., who heads GM's advanced-technology-vehicles program. "I think the concept makes the most sense in commercial buses and urban-delivery vehicles, where there's a duty cycle that includes a lot of stop-and-go driving."

Purcell might get an argument from today's suburban commuter, whose congested route means stop-and-go traffic is the norm, whether it's on a superhighway or not. Hybrid technology may also be perfect for big sport-utility and military vehicles, which could become much more fuel efficient with hybrid drivetrains. It's not surprising that the military is heavily invested in all kinds of alternative transportation research, since fuel cells and hybrids offer the potential of long-range, multi-fueled vehicles with little noise or heat signature to give away their position.

DO DIESEL HYBRIDS HAVE A FUTURE?

There are no outstanding technical hurdles with hybrids, and they stand the best chance of meeting the fuel-efficiency goals of the joint venture between government and industry known as PNGV, the Partnership for a New Generation of Vehicles (see chapter 8). But hybrids are most definitely *not* zero-emission. And what's more, many of them, like the GM parallel hybrid and the Chrysler ESX2, are diesels. They're very clean, lean-burning, and efficient diesels, of course, but still kissing cousins of those dirty, smelly buses that most polluted cities are trying to purge from their streets.

The PNGV goal, a major commitment by the auto industry, is to build an eighty-miles-per-gallon prototype car by 2004. But is the consortium willing to compromise on emissions to meet its fuel-economy target? And just how dirty are these direct-injection diesels, anyway?

"Things are getting *very* interesting," says Roland Hwang, the Union of Concerned Scientists' point man on auto emissions. "Several auto companies have tried to sell me on diesel as a fuel future pathway, because it's so compatible with what we have now." But Hwang thinks that current diesel technology is "probably about five years behind what the air regulators want by 2004." He points out that diesel engines account for nearly half of U.S. emissions of smog-producing nitrogen oxides and more than two-thirds of its particulate matter. Sooty particulates released directly into the air and easily inhaled irritate the eyes and nose and aggravate respiratory problems (including the asthma that afflicts thirteen million Americans).[16] The EPA, in a draft report, blamed diesel exhaust for an increased lung-cancer rate, as well as for such relatively minor side effects as light-headedness, numbness, and vomiting.[17] The California Air Resources Board, whose approval is vital if the automakers are to sell their hybrid cars nationwide, turned away fierce trucking industry lobbying to declare, in the summer of 1998, that particulate matter contained in diesel exhaust is a "toxic air contaminant."

Hwang still thinks the auto companies may be able to win approval for their advanced diesels as "clean" technology. The Clinton administration, for one, has been a notable supporter of diesel research, through the industry-government PNGV. But the scientists at the National Research Council, which reviews PNGV's work, are somewhat skeptical that direct-injection diesels can be cleaned up enough to meet emission standards. Still, Hwang thinks the companies can combine rapid technical improvement with the same exploitation of regulatory loopholes that allows them to produce spiraling numbers of gas-guzzling sport-utilities without paying any federal fuel economy penalties. "But that would lock us into the fossil-fuel infrastructure for another several decades," Hwang says. The Union of Concerned Scientists would rather see the carmakers leapfrog right past hybrids and go directly to fuel cells, but this seems unlikely given the remaining challenges. Unsurprisingly, the PNGV cars shown off by both GM and Ford were diesel hybrids.

As the twentieth century ends, there are many quite different technologies competing for a chance at the marketplace. A number of these possible solutions to our automotive pollution problems will fall by the wayside. I don't really think the flywheel will fly, though I could be proven wrong. But both fuel-cell and hybrid cars, revolutionary approaches to transportation that are the subject of billions of dollars of industry R&D, are reaching the consumer. As with early gasoline cars, there are bound to be serious teething problems—but actually getting the vehicles into production is a serious milestone. Reading about the technology in these cars, and seeing them on test tracks and show stands, I *wanted* them to work. But will these exotic and complicated cars stand up to the rigors of daily life? It's too early to say, but evidence *is* beginning to roll in on the first commercial battery EVs.

CHAPTER 4

ROAD WARRIORS AND EARLY ADOPTERS: LIVING WITH A BATTERY-POWERED EV

IT WAS HARD to miss the shiny bronze metallic EV Plus at Paragon House of Honda in Woodside, Queens. The electric car was right by the front door, for one thing, and for another the wall was decorated with a giant blowup of the dealership's full-page *New York Times* ad. "A Car with a Cord," it said. "Sounds Like a Honda."

In the summer of 1998, Paragon House of Honda became the first dealer in New York, and, actually, the first on the East Coast, to sell the EV Plus, a nickel metal hydride battery commuter car that looks something like a streamlined Civic. The EV Plus was no road rocket like the EV1—it took 17.7 seconds to reach sixty miles per hour—but its advanced batteries gave it an impressive 125-mile range. (In the real world, owners said, seventy to one hundred miles was more like it.) The EV Plus was also a four-seater, a fact that made it popular with families.

The EV Plus was undeniably attractive, but operating an electric car in New York is a daunting prospect. While, in 1901, New York City had a central charging station that could pump up one thousand EV batteries at a time,[1] today the great metropolis has only a few scattered electric utilities so equipped. New York State had no strong grassroots lobbying effort for clean cars and no quasi-state agency, such as California's CALSTART, cheerleading for them. What's more, modern battery makers say their products just can't cope with the state's winter chill.

Despite these unpromising portents, then Governor Mario Cuomo stuck his neck out in 1991, using the just-passed federal Clean Air Act amendments to get tough with the car companies. Emulating California, he required that 2 percent of all the vehicles sold in the state beginning in 1998 be electric. California rescinded its 2 percent mandate in 1996, though it kept the much more onerous requirement for 10 percent by 2003. But New York, now governed by an unlikely environmentalist named George Pataki, stuck with the 1998 deadline, provoking rage—and a lawsuit—from the American Automobile Manufacturers Association, whose members were the Big Three.

Carmakers say again and again that innovation, not regulation, is what motivates them, and the New York mandate particularly rankled. The state had no EV infrastructure, carmakers said, but it was trying to legislate that the auto industry create one. Further, New York stood to profiteer from the large fines it would levy on carmakers that missed their deadlines. "There has never been anything in American history of similar government interference in the supply side of a major industry," said a rather dramatic report commissioned by the AAMA.[2]

Despite their loud protests, the automakers slowly and grudgingly began to build a dealer and service base in New York. By mid-1998, Ford had eight such dealerships, from Great Neck to Rochester. Momentum began to build.

The state's law was initially upheld. But in August 1998, the U.S. Court of Appeals sided with the AAMA and struck it down, saying that because of the wording in the interstate agreement, New York was legally bound to follow California. Most of the carmakers that had begun to establish fledgling EV dealer networks in New York promptly abandoned the effort (though Ford remained).

But that wasn't the last word, either. In November of 1999, Governor George Pataki tried another approach. Instead of dictating Empire State EVs, he said New York would adopt California's stringent auto emissions standards, an order with a potentially huge impact on the car industry—even beyond the state's borders. Massachusetts announced it would sign on also, joining three very large auto markets in a common purpose. The states had, in effect, forced the industry to accept a new national emissions standard. Did all these machinations faze people like Chris Howard, sales manager at Paragon House of Honda? Not at all. Speaking on the very day that the legal decision was handed down, Howard said that the company's EV program had too much internal momentum to be stopped by court orders. "Considering the investment Honda has in EVs, I'd say they're in it for the long-term," Howard informed me. In a month and a half, he'd taken fifty people on test rides, with more than half showing strong interest and ten becoming "driver candidates," people willing to pay $445 for a monthly lease and a one-time fee of $750 for a home charger.

Unfortunately, the early momentum on the EV Plus was not maintained, and Honda leased only a few hundred over two years. In the spring of 1999, Honda pulled the plug on the battery car, obviously preferring to concentrate on the hybrid Insight and its ongoing fuel-cell efforts. It was a definite signal that carmakers were becoming disillusioned with the limited range and slow consumer response to battery-powered electrics. The EV Plus was very popular with the people who took it home, but there weren't enough of them. "We're highly disappointed in Honda," said Tom Cackette of the California

Air Resources Board when the Japanese automaker made the decision public. Honda is likely to try to meet California's 2003 mandate with a combination of superclean internal-combustion cars, hybrids like the Insight, and fuel cells (if they're ready).

EARLY ADOPTERS

It may take a decade for fuel-cell cars to reach the public in any numbers, but production battery EVs started to trickle out of the big carmakers in the mid-1990s. And who is leasing them? A pattern is emerging in these "early adopters." They're upper-middle-class achievers. They have high-tech jobs. Some, not all, care a great deal about the environment. The biggest thing they have in common is their love for their EVs. These people, the only Americans in living memory to routinely drive electric cars, are loyal to a degree unknown in the car business.

"The numbers are small but the dedication of these customers is beyond belief," says GM's Robert Purcell Jr., whose Advanced Technology Vehicles division is the brain trust behind the EV1. "I've spent time with Ferrari and Lamborghini owners, and never saw this level of enthusiasm. These are people on a personal mission. They believe they are pioneers bringing in the next generation of the car business. They're just the customers we want, high-energy people carrying the message."

Many of the battery EV's initial customers were strong environmentalists, asking pointed questions not only about global warming and urban smog but also about charging times, nickel metal hydride batteries, and just what is meant by "zero emissions." It's an educated, probing customer base, one typical of what are called early adopters, people who get out in front on everything from cell phones to DVD players. Colin Summers, a California EV1 driver, certainly fits the mold. "I had a laptop when people didn't have desktops," he says. "I carried a nationwide pager when a lot of people didn't know how pagers worked. I had a cell phone . . . when most [of my] friends didn't have cordless phones yet. So I wanted to be in on this technology, too."[3]

Honda introduced the EV Plus in California in 1997, not long after GM's EV1 made its debut in the same state. According to Robert Beinenfeld, a senior manager in Honda's alternative fuel vehicles division, Honda's research showed the company that it couldn't successfully market a car with lead-acid batteries, because of range restrictions. Instead, Honda became the first volume automaker to go forward with the more expensive nickel metal hydride packs. Although seventy to one hundred miles is still not a lot of range, many members of the tiny EV Plus fraternity said they were well pleased with the car's versatility.

Tim Hastrup of Granite Bay, northeast of Sacramento, is a typical commuter in that his morning ride (to high-tech employer Hewlett-Packard, where two co-workers also drive EVs) is only ten miles. "It meets ninety-five percent of our family's car needs," Hastrup says. "We still have a gas car, but for most of our day-to-day needs we take the EV." Like most EV drivers, he's an environmentalist. "I can't see how we can continue to burn petroleum as wastefully as we do," he says. In fact, Hastrup, whose family of four fits comfortably in the car, "likes everything about the EV Plus but the price." He had heard—correctly—that replacement battery packs can cost $20,000. Hastrup would have gladly bought his car, even for $30,000, if there had been a purchase option.

Pete Lord of Santa Clarita, California, who shares his EV Plus with his wife and teenage daughter, says he's gotten as much as ninety-two miles out of a charge. He's a confirmed early adopter. "I like the newness of it," admits Lord, a biomedical engineer. Like many EV lessees, he's held on to his gas car, but he finds that the precaution isn't necessary. "Our van sits in front of the house as a monument to overly cautious thinking," he said, speaking before Honda killed its EV. Like other EV Plus lessees, Lord was suddenly faced with the worry that Honda would no longer provide an infrastructure for his orphaned car.

The GM EV1 has even more enthusiastic owners. They've formed a club, whose membership includes such entertainment industry figures as Danny DeVito, Ted Danson, Robin Williams, Kevin Nealon, and Alexandra Paul. Perhaps the dean of EV1 owners is actor Ed Begley Jr., who actually goes so far as to avoid *riding* in a gas-powered car. "I've been driving electrics on and off since 1970, and exclusively since 1990, but nothing I've tried is anything like the EV1," he says.

Marvin Rush, a cinematographer, is even more enthusiastic. In his first year and a half, he's put twenty-six thousand miles on his EV1 and loves "crucifying gas cars that think this is a golf cart. With a zero-to-sixty time of less than nine seconds, I leave them gasping in my dust." Rush has actually gone so far as to create and pay for EV1 print and radio advertising on his own. "Blow the doors off your neighbor's car while you clean the air on your way to work," reads one spot. "EV1 . . . It's Not What You Think!"

Rush, a co-founder of the EV1 Club, is passionate not only about the car but also about the company that made it. "The California electric car mandates happened because GM took a chance on the EV1," he says. "GM is moving away from internal combustion. I believe that big things are done by big companies, and it will take corporate clout to get us away from piston engines. This is beyond left and right, beyond politics. Global warming is real, and it will kick our ass unless we do something about it."

Fred Beer, a semiretired Southern California Edison numbers cruncher who lives down the freeway from Rush in Oceanside, is also in the EV1 Club, one of seventy-five owners who live within driving distance of each other. It's a tight-knit group. They socialize a lot, exchange daily e-mails, compare notes about range.

Beer's EV1 saga began with a car crash. "I totaled my Corvette in the rain," he says. "I had two weeks to get another car, and Edison offered me a subsidy on the EV1. It was timed perfectly." Beer periodically works as a consultant in San Diego, fifty miles away. A round-trip is beyond his lead-acid EV1, so he sometimes parks the car at the San Diego Convention Center. The public charger there can recharge the car while Beer works. From there, he can take the city's new trolley system to his office. "It's an all-electric ride," Beer says. It's also a model for the low-emissions commute of the future, though this kind of synergy is available today in only a few places.

Beer admits he's "not that involved in the ecology thing," but he's worried that we're running out of fossil fuels. And he doesn't believe he's had to give up anything to drive an EV1. "The performance has been excellent. I was initially taken aback by the acceleration. This car can beat any car zero to sixty. It takes off like a jet—I feel the pressure in my back." Beer has also driven his EV1 to Los Angeles with a map of charge stations in his lap. "We stopped and had breakfast while we were charging," he says.

Tom Dowling, a Sacramento-based bank technology manager, takes his EV1 even farther. "I drove six hundred miles last week," he says. For his trip to Lake Almanor, in northern California, he took along his charger and plugged in anywhere he could find the proper voltage. "I searched out places that had 220 current," he explains, "welding stations, RV suppliers, even somebody's 220-volt dryer outlet. I knew I would need a charge in Belden on the Feather River, population ten, so I got on the Internet and tracked down one of the only businesses there. The owner was only too happy to help me, and he had a 220 welder. The irony was that his gas pumps had a sign saying, 'No gas today.' But I got my charge." Dowling admits to some close calls. At one point he was twenty-nine miles from his destination with thirteen miles left on his charge, but the road headed downhill, and the EV1's motor turned into a generator, getting him there with miles to spare.

Dowling's distance adventures pale before those of Kris Trexler, a writer for many TV shows. Trexler is so nuts about his EV1 that he drove it 3,275 miles, from Los Angeles to its hometown of Detroit, on Charge Across America, a trip that took three weeks. Like Dowling, he yanked the MagneCharger out of his garage, put it on wheels, and went off in search of 220 current. Trexler had a wonderful time, never broke down or ran out of power, and arrived in Troy, Michigan, in June 1998 to a hero's welcome. The trip was in rather sharp contrast to the aborted journey across America made by writer Noel Perrin and documented in his book *Solo.* Perrin's Solar Electric Ford Escort saw most of the U.S.A. from the back of a flatbed truck.

In a Hollywood that seems besotted with EVs, Begley stands out as particularly committed to them. He's put seventeen thousand miles on his EV1, which he takes out only when his bicycle won't do and public transportation isn't available. A frequent public speaker about clean transportation, Begley used to think that the auto companies were conspiring to keep EVs off the market, and he fought long, hard, and successfully for the state law that requires the industry to build and sell them. Now he thinks that competition has become the driving force behind a vibrant EV movement. "I know actors," he says, "and these car executives aren't acting when they tell me they're committed to making this work. I screamed for the mandates, but now I think EVs would happen even without them."

Begley got his first EV the year of the first Earth Day, 1970. It was a slow and heavy Taylor-Dunn, a glorified golf cart. "I was fed up with the smog," he says. "I grew up in L.A., and I couldn't run to the mailbox without wheezing, and I'm not even an asthmatic." A series of good, bad, and indifferent EVs followed the Taylor-Dunn, but Begley was temporarily side-tracked. "A friend told me, 'Ed, you're a dummy. You're going to all this trouble and still generating the same amount of pollution because of the power plants that produce your electricity.' I believed that then, but I later found out that, when everything is accounted for, an EV is actually responsible for only ten percent of the pollution produced by a gasoline car."

These days, Begley wonders why more consumers aren't stepping up to lease an EV1, whose monthly price is no higher than that of a BMW or Lexus. "There's no nefarious conspiracy keeping people away from the EV dealerships," he says. "People who call themselves environmentalists should put their money down, because cars like the EV1 and the EV Plus can handle ninety percent of the trips they need to make."

Ford's electric Ranger, being a working truck and not an undeniably sexy sports car, doesn't inspire quite the same loyalty as the EV1. Robert Gerber runs Gerber Electronics, a seventy-employee distributor of industrial electronics in Norwood, Massachusetts. The Ranger isn't a cause for him, but he says he's had a positive experience. "The truck accelerates beautifully, handles well, stops well, and is ghostly quiet," he says. Gerber would change a few things, though. He's not sure why the truck needs power-robbing air-conditioning. He'd love to see an emergency charging system that could simply plug into household current. And that lead-acid battery pack is a definite hindrance. "The fifty-mile range limits where you can go," says the veteran of more than one near-empty-battery scare. "But for running around in a ten-mile radius of the plant, it's great."

Gerber is no armchair environmentalist. He's a sponsor of the MIT Solar Car Challenge, and his company also runs a bi-fuel (gasoline–natural gas) Ford Contour, which tanks up at a high-pressure Boston Gas facility just down the road from his office. The Contour has the Ranger's range problem beat—it can actually go a bit farther than a gas-engined Contour—but it can't claim to be a zero-emission vehicle.

Gerber's concerns about the lead-acid batteries were addressed in late 1998, when Ford first made one hundred nickel metal hydride versions of the 1999 Ranger available to California customers. "A lot of my utility customers wouldn't talk to me about lead-acid trucks," says Ray Roy, EV fleet sales manager at Senator Ford in Sacramento. "They really wanted a one-hundred-mile range, which the nickel metal hydride version can deliver."

The Toyota Prius hybrid car can deliver much more than that, which is one reason its sales in Japan have dwarfed that of other EVs. It has an unbelievable range of eight hundred miles, and its sixty-six-miles-per-gallon fuel economy is frequently cited by owners as their reason for buying it. Shizuma Hisashi, for instance, a twenty-three-year-old Tokyo college student, says environmental concern was "a strong reason for my purchase." He loves the car and thinks hybrid drivetrains will replace internal-combustion engines, but he admits that the regenerative brakes, which salvage stopping energy to keep the batteries charged, "feel kind of strange." (Actually, they *are* strange, in that it's a brake-by-wire system with no mechanical connection between pedal and wheel, and the feel varies considerably, depending on the amount of regeneration.)

Watanabe Haruka is a Tokyo company owner who loves golf and his Prius, in that order. "This is the car for the twenty-first century," he proclaims. "It's futuristic and stands out against conventional cars. And women think it's cute." He puts eight hundred miles a month on the car, mostly for business trips. Kawahara Teruko, also of Tokyo, uses her car for shopping and errands, racking up one thousand miles a month. The sixty-eight-year-old homemaker, who agrees that the regen brakes are a little odd, got her car "to be kind to the earth."

BACKYARD BUILDERS AND SHADE-TREE MECHANICS

It's hard to find an EV driver who *doesn't* love his car. Even more fanatic than the people with leased EVs such as the Ranger and EV1 are the backyard tinkerers who put their own cars together. As production EVs become readily available, do-it-yourselfers are becoming less common, but they are a dedicated lot.[4] John Stockberger of Chicago, for instance, converted an old Pinto using a surplus military aircraft generator and a burned-out shell he found in a wrecking yard. Californian Bill Palmer used golf-cart batteries to give his old Chevy an eighty-mile range. And Clarence Ellers's Aztec 7 XE, a futuristic sports car with lift-up gull-wing doors, has served as a driving advertisement for the owner's conversion manual.

Ron Kaylor Jr., an electrical engineer from Menlo Park, California, began building electric cars in the early 1960s; he specialized in VW Beetle conversions (using motors that originated in F-100 fighter planes) and, in the early 1970s, offered his own Kaylor Hybrid Module that could give the Bug a four-hundred-mile range.

Not many people have built their own hybrids, because they are considerably more complex than simple battery EVs. But in 1979, Dave Arthurs of Springdale, Arkansas, spent $1,500 turning a garden-variety Opel GT into a hybrid that could get seventy-five miles per gallon, using a six-horsepower lawn-mower engine, a four-hundred-amp electric motor, and an array of six-volt batteries. Nobody would build a car like this today, because lawn-mower engines are notoriously dirty (not to mention short-lived), but Arthurs's car so impressed the editors of *Mother Earth News* that they built their own hybrid, using a 1973 Subaru as the base. That car averaged 83.6 miles per gallon. Memories of the 1973 Arab oil embargo were still fresh, so when *Mother Earth News* published the plans, sixty thousand readers wrote in for them.[5] Arthurs continued at it in the 1990s, building a ninety-miles-per-gallon Toyota pickup with a nine-horsepower diesel engine, and experimenting with a rotary engine that would run on peanut oil.

THE PRIVATEERS: EV MAKERS ON THEIR OWN

If it was easy to start a car company, instead of a Big Three we'd have a Big Three Hundred. That was indeed the case before the Depression, when automakers McDonald, McFarlan, Mercer, Meteor, Mitchell, Monitor, Monroe, Moon, and Moore were all going concerns (and that's just the Ms).

A huge drop in American discretionary income during the Depression years killed these and hundreds of other small manufacturers. Most simply lacked the capitalization to weather the storm. But even after World War II, when pent-up demand was released and owning a car became an attainable dream for most middle-class families, the independents found it hard to compete. Hudson, for instance, which had managed to survive the war years and bring out its spectacularly advanced "step-down" cars in 1948, found itself unable to update the model quickly enough for consumers entranced with the latest from Ford, Chrysler, and General Motors. The hideously ugly '57 Hornet was the end of the line.

Tiny Hudson, however, was a giant compared to the start-up EV companies that began to appear on the American scene in the wake of the 1973 Arab oil embargo. Linear Alpha produced conversions, as did Electric Fuel Propulsion, whose one-hundred-mile-range Transformer had a brief moment in the sun. Sebring-Vanguard, maker of the CitiCar, was able to sell all the EVs it could make for a short time, but customer loyalty proved fickle when pump prices came back down and the gas lines disappeared.

None of these companies was likely to set Detroit on its ear. The outfit that seemed to have the necessary moxie was California-based U.S. Electricar, which had its first success in the 1970s converting the tiny Renault LeCar to EV status. By 1994, this once-sleepy small company had undergone a makeover: it had begun to act like a conglomerate-in-training, with ultramodern offices all over California and in Singapore, too. The publicly traded company had three hundred workers and three plants with 170,000 square feet of manufacturing space. Dave Brandmeyer, the aggressive vice president I met in Los Angeles that year, said that he'd just come from a meeting with GM chairman Jack Smith, who "wants to work with us to build thousands of vehicles and meet the [1998 California] mandates." Indeed, both Ford and GM did come calling, but no deals resulted.

U.S. Electricar even had a conscience. Its Los Angeles factory opened in 1993 in the Watts Enterprise Zone. The several dozen employees were mainly long-term-unemployed blacks and Mexicans working through the state Joint Training Partnership to convert Geo Prizms and Chevy S-10 pickups to EVs. "Part of our battle is not to sell cars but to lift people's consciousness," the plant's African-American manager, Rodney Maddox, told me. "Electric cars have brought jobs to this community, where it's hard to find any kind of job."

But the company that seemed to have everything couldn't sustain its forward momentum. Even as it was receiving ecstatic press notices in 1994, it was expanding too far, too fast.

In March 1995, rumors about impending disaster were borne out when the company announced large losses, and it was revealed that top officials of the company had been jettisoning their own stock.[6] When U.S. Electricar folded, company bosses said the Republican landslide of 1994 had done them in by chilling the atmosphere for federal EV subsidies, but customers complained of receiving blatantly defective cars.

U.S. Electricar also had some strange ways of doing business. While hinting about coming big deals in Detroit, the company was also openly contemptuous of the Big Three, and so self-confident that it didn't think it would need an industry partner. Like some Internet companies today, it expanded rapidly without any solid base of sales.

"They ramped the company up in their garage, but didn't do anything else right," says James Worden, head of Electricar's onetime East Coast rival, Massachusetts-based Solectria. "They burned up money so fast they were offering Learjet rides when it was quicker to drive. It made a real impression, and even utility guys bought stock, but the cars were not well engineered. They kept pushing the idea that the Big Three were evil, but you don't bad-mouth Ford and GM when you're dependent on them for parts and supplies."

Solectria had more staying power. By Wall Street standards, its small-scale beginnings in 1989 were not auspicious. Founder Worden had just graduated from the Massachusetts Institute of Technology, where he'd spent his time with the Solar Car Racing Team, setting EV speed records. No one doubted his bravery—he almost burned up in a racetrack battery fire in 1992—but his entrepreneurial instincts weren't immediately apparent.

The company started out selling parts—solar panels, electric motor controllers, and converters—to other college EV racing teams. Parts are still the mainstay of Solectria's business, and the steady demand for them is one of the main reasons the company has stayed in operation for a decade. The parts business soon led to full-fledged EV conversions on compact cars and pickup trucks. Because even the most basic conversion sells for $33,000, most sales have been to utilities and government agencies.

The other major reason Solectria has endured is government funding. Contracts from the Defense Advanced Research Projects Agency (DARPA), the military's technology incubator (see chapter 8), floated Solectria's bid for the big time, the lightweight Sunrise EV. At last, Worden said, Solectria was going to build its own battery car from the ground up, not just make conversions. The little company would become a full-fledged carmaker, and it began shopping for additional factory space in Massachusetts and Connecticut.

Solectria got a lot of notice when it boasted confidently, in 1994, that it would soon be mass-producing twenty thousand Sunrise cars a year. No independent EV maker has ever had that kind of volume, and Solectria wasn't likely to achieve it without a major automaker partner. Unfortunately, Solectria was knocking on Big Three doors just as the companies' own EV programs were being launched. Although the Sunrise was an innovative design, it was nothing Detroit couldn't do on its own (and, in fact, did soon after with the GM EV1). The Sunrise didn't happen; only a few prototypes were built.

Today, the company, which is housed in the kind of industrial park whose streets are named Progress Avenue and Enterprise Boulevard, turns out the Force EV, a converted Chevrolet Metro, and has set its sights on something rather less momentous than a mass-market electric automobile: the practical GM-based CitiVan, an urban delivery vehicle. The carcasses of several Sunrises sit on a rack in the back parking lot.

Economies of scale also plague other independent EV companies, like Unique Mobility (briefly in the news in 1998, due to former Mouseketeer Darlene Fay Gillespie's bizarre manipulation of the company's stock), conversion specialists like Green Motorworks and the Solar Car Corporation, and many other small, ambitious companies.

Some of these start-ups have become post-retirement hobbies for onetime Detroit kingpins. Former Chrysler chairman Lee Iacocca raised $10 million to start EV Global, a company that makes electric bikes and other EVs. EV Global buys batteries from Energy Conversion Devices, where former GM chairman Robert Stempel serves as senior consultant and frequent public spokesman. The two have become friends and business partners in a way they never were when they ran rival auto giants.

Major marquee names like Iacocca's look good on letterhead, but so far they haven't added many other benefits to the cash-strapped independent EV industry, where learning to innovate with limited resources may matter more than a gilt-edged corporate résumé. Some of the best EV designers have been creative independents. Paul MacCready, for example, whose brilliant breakthroughs with lightweight race cars made the EV1 possible, worked not in Detroit but in his own small California R&D firm, AeroVironment.

But even if a Detroit pedigree is a liability in the hothouse, Silicon Valley–like atmosphere of the small firms, the industry's backing has proven vital to the nuts-and-bolts process of actually getting a car to market at a price customers are willing to pay. This proved much harder than it at first appeared to many hopeful entrepreneurs, who thought they could beat the Big Three at their own game. How could Solectria, for instance, sell its Force EV at a consumer-friendly price, when the car was based on a fully equipped Chevrolet Metro that the company had to buy at retail prices off a dealer's lot? It wasn't until 1998, seven years into building the Force, that Solectria was finally able to buy engineless "glider" cars from GM.

By the late 1990s, it was clear that only the big automakers themselves, with the largest cash reserves in global industry, could make the EV revolution happen. Ironically, these were the very same companies that had tried to kill electric cars through most of the previous decade.

MARKET CHALLENGES

There are 200 million cars in the United States, but only about 3,500 of them are "freeway-capable" EVs, the majority home-made conversions. When a sales manager at a major New England dealership was asked if he ever marketed his electric trucks to individuals, as opposed to utility-based fleets, he laughed. "What for? Who would want a unit that can only go fifty miles?"

It's not surprising, then, that EVs have been an afterthought for the auto industry and that a serious marketing plan is only now emerging, as the technology improves and the 2003 California mandates approach. EVs, once a footnote in carmakers' annual reports, now merit their own four-color publications. GM even produces a kids' book called *Daniel and His Electric Car.*

The numbers are still tiny. Sales and lease figures for battery EVs have to be qualified by the fact that the cars have been marketed in only a few states, with marginal advertising. But even by that measure, they've failed to perform saleswise. Honda's decision to discontinue the EV Plus program was disappointing but not that surprising to many EV partisans. By the spring of 1999, after three years on the market, GM had leased only 660 EV1s and 500 S-10 pickup electrics. Toyota's RAV4 EV, on the market only since the end of 1997, was doing comparatively better with 507 vehicles out the door. Ford moved only 445 Ford Ranger pickups, and only 267 people had signed leases on the EV Plus. In its entire ten-year history as an EV maker, Solectria has sold only 350 converted cars and trucks. A recall of the whole line would fit in a supermarket parking lot.

But Honda's Robert Beinenfeld thinks the numbers may not tell the story. He calls the EV Plus a success, despite its short life span, because it produced data from a wide customer base. "Only Honda and GM have been involved in the retail market at this point," he said. "If you stay away from the public, and only lease your vehicles in fleets to utility-meter readers, you're protecting yourself from risk. There's one customer—the utility—instead of fifty individual ones." Indeed, Susan Romeo, director of marketing for California's CALSTART consortium, thinks the lease numbers would be a lot higher if the companies simply made more cars available.

Could it be that Americans just don't like EVs? That's the line in some industry publications, but it's not borne out by opinion polls, which routinely show that the motorists' love affair is with the private automobile, not solely with its internal-combustion form. A healthy minority would trade in their current car for an EV, provided it could perform as well and not cost any more. A poll of California new car buyers conducted by the University of California at Davis in 1995 found that almost half would prefer an EV over a gasoline car. But they want their EV to cover three hundred miles on a single charge and be reasonably priced. And that caveat has encouraged the sale of a million trucks and SUVs. There is no such EV currently on the market.

Range is, obviously, the main problem for most potential buyers. Though EV advocates point out that most commuters take round-trips of fifty miles or less, the distance limitation is an important one. Most SUV buyers never go off road, but the *potential* for that kind of excursion is what gets them to write out the check. An added sore point is that the luxury accessories Americans love—air-conditioning, power windows and locks, fancy stereo systems—depend on power from the onboard batteries, and so sap the car's range even further. Conventional bare-bones compacts aren't moving, so vanilla EVs are an inherently even harder sell in the marketplace.

Price is also a drawback. Because they're built in such limited numbers and because the materials that go into them (not to mention their battery packs, which in the case of the lithium-ion cells in Nissan's Altra EV cost in the six-figure range) are so expensive, EVs sell for $30,000 and up. Automakers offer lease deals so they can hide the real cost of the vehicle and also insulate the customer from the uncertainty of a looming and costly battery-pack replacement, but even these subsidized leases require an extra $100 or so in monthly payments. Leasing allows the manufacturers to "retain control" of the EV, as Mike Clement, Chrysler's manager of alternative vehicle sales and marketing, describes it. "The technology is changing very fast," he says, "and we don't want customers to have a two- or three-year-old vehicle that is out of warranty, with obsolescent, very expensive parts." The strategy makes sense, but so far it hasn't worked very well. Car buyers seem willing to purchase or lease EVs in the abstract, but can get scared off by the price premium and the endemic uncertainties, including the likely shortage of charging stations.

It's reasonable to conclude from all this that consumers don't care about the environment. But there you'd be wrong. Almost half of all adult Americans dutifully recycle their cans and bottles.[7] In 1995, 27 percent of U.S. voters made their closed-curtain decisions solely on the basis of candidates' track records on environmental issues. A clear majority identify themselves as environmentalists and believe that current environmental laws don't go far enough. Fifty percent, according to the 1996 Roper "Green Gauge," worry that auto air pollution is a "very serious" problem.[8] There's obviously a good potential EV customer base there, but to turn sympathetic tire kickers into lessees, or, eventually, owners, the product has to be very good indeed. And until very recently, it wasn't.

TEST-DRIVING EVS

I should confess here that I've never built or owned an electric car, though I've driven dozens of them, beginning in around 1994, when the job of editing *E Magazine* came my way and I began to educate myself about environmental issues. (It was disconcerting, to say the least, to learn that my hobby of collecting classic cars and my growing concern for the environment didn't necessarily mesh.)

In the summer of 1994, on assignment for *E,* I went to southern California to see for myself what was then becoming a national incubator for a nascent EV industry. The cars then reflected the amount of work that went into them and ranged from stunningly futuristic to embarrassingly jury-rigged. Definitely one of the latter was the Danish-made Kewet El-Jet I encountered at Green Motorworks in North Hollywood. At $18,350 fully loaded, the El-Jet certainly had the virtue of affordability in an era when most EVs started at $25,000. But that's about it for the plus column. For the price of a Taurus, the environmentally conscious buyer got a square and squat fiberglass box, powered by lead-acid batteries. White ones looked like enclosed golf carts.

Driving the Kewet would be a familiar experience to anyone familiar with either the Yugoslavian Yugo or the East German Trabant. The car was slow and noisy and had lackluster attention to detail, summed up by primitive sliding, rather than wind-up, side windows. While most EVs are almost preternaturally quiet, this one offered a symphony of road noise and rattles.

The Kewet was blessed with a huge front window, but that didn't mean I could see out of it. There were torrential rains in southern California that day, and the Kewet was as steamy as a Madonna video. My Green Motorworks co-pilot kept assuring me that the defroster would kick in, but we made it back to base only by peering through a small hole created by frequent hand wipes.

On that same trip to California, I also drove much better cars. Southern California Edison lent me a Ford Ecostar van, the company's first serious EV contender since Thomas Edison inspired Henry Ford. The Ecostar was the first EV I'd seen that looked like an actual production car, instead of a hasty conversion. Its purpose-built recharge port, its integrated dashboard battery meter, and its fully finished appearance all suggested the work of a major automaker.

When EVs are as well conceived as the Ecostar, driving them isn't necessarily an "experience" to compare with a cruise in a Kewet. It's fun, but not a dramatic departure. After half an hour, you forget you're in an electric car. EV makers, not wanting to confuse the public with the shock of the new, often design them to be as "transparent" to regular motorists as possible. That means controls are just where you'd expect them to be, and the "futuristic" details so often seen in show cars are kept to a minimum.

The Ecostar used high-tech and high-temperature sodium-sulfur batteries, which were seen as very promising because they offered decent range. With them, the van could sprint from zero to sixty in twelve seconds. Steering it into ferocious Los Angeles traffic, I kept thinking the van had stalled, because there was no engine noise. (EVs are at their most remarkable when simply stopped at a traffic light, because they sit there doing nothing at all.) The Ecostar had no trouble keeping up with its fossil-fueled brethren, and it felt rock solid, not like a developmental prototype but ready for commuting duty right then and there.

Unfortunately, Ford never built more than about eighty Ecostars. Those sodium-sulfur batteries, while offering the tantalizing possibility of five times the range of lead-acid, proved temperamental to say the least. The batteries operated at the incredible temperature of 500 degrees Fahrenheit and were very sensitive to cold. Even worse, they were dangerous. The program was terminated after two California demonstrators caught fire in separate incidents soon after my test ride. It was further proof that battery packs are both the strength and the Achilles' heel of zero-emission EVs.

U.S. Electricar was very much a going concern then, and I drove its lead-acid Electricar Prizm in Torrance at Hughes Power Control Systems, the GM subsidiary that built the car's DC-to-AC inverter. Like Ford, Electricar knew that Americans didn't want their EVs to be exercises in space-age futurism. There were few clues, other than a gas gauge converted to a range meter, that the Prizm was anything other than a standard production model. Even the batteries were hidden, in a covered tunnel underneath the car.

Driving the Prizm gave the game away immediately. Most EVs have fantastic low-end torque, translating into respectable zero-to-sixty times, but the Prizm was sluggish off the line. Acceleration was slow, but it did build relentlessly. Seeing as its Hughes custodian had no objection, I soon had the car up to eighty miles per hour, a very quiet eighty. Hughes's Fred Silver showed me the car's innovative recharging paddle, a unit so safe that the company demonstrated it by throwing it into a swimming pool. I was impressed.

The California test rides were fine, but I wanted more of a chance to drive an EV in real-world situations. I'm a syndicated car columnist, so I'm used to having cars loaned to me by their manufacturers for one-week periods, but this proved nearly impossible to do with an EV. First, many of the companies were in California and I was in Connecticut, and second, most of the available cars were what amounted to $100,000 prototypes, affordable only to big utility companies. But I finally persuaded U.S. Electricar that I was worthy of their trust. After weeks of delay, an Electricar Prizm arrived on a flatbed truck.

Because there's no ground-level power in the two-story office building where I work and the batteries appeared to be depleted, I brought a one-hundred-foot extension cord from home, draped it out of the window, and plugged it into the car's socket. We left it connected all through the work day, but at 5:00 P.M. it still seemed to be only partially charged.

Thinking that a half-full battery would get me the fifteen miles to my house, I set out confidently. The car performed much as it had done in California, painfully slow at first but gradually gaining momentum to cruise at comfortable highway speeds. Five miles from home, I relaxed enough to flip on the radio and scan for some jazz. But without much warning, the car suddenly started to slow down, even though ample charge still showed on the gauge. I was soon moving down I-95 at twenty miles per hour. Luckily, an exit appeared and I coasted down it, realizing I was only about half a mile from a friend's beach house.

By now the car, with the pedal to the metal, was moving at about ten miles per hour, the gauge still cheerfully indicating a third of a charge. Accompanied by a chorus of honking horns, I barely made my friend's driveway before the car conked out completely.

Nobody was home.

I considered my options and found I didn't have many. My wife was at a business meeting and couldn't come get me. It was too far to walk. Then I noticed the porch had an overhead light, and I had all the hardware to hook into it. I plugged in and was satisfied to see the little red charging indicator come on. Thinking all would be well, I decided to leave a note and the car, which I now knew needed many hours of charging.

That night I got the horrifying news. My friend's mother had arrived home to find her porch smoldering and the ceiling around the light blackened. The extension cord was nearly red hot. Dumb luck prevented the place from burning down completely. I was soon informed by an incredulous U.S. Electricar that EVs need to be plugged into grounded outlets, not normally found in houses with fifty-year-old wiring. But they couldn't tell me why the car had failed in the first place.

I paid to fix the fire-damaged porch. And I convinced U.S. Electricar to give me another car. It proved as troublesome as the first one, and after it was ignominiously towed away, I never heard from U.S. Electricar again. Soon after, the company left the U.S. market.

The Electricar debacle, in which the company had expanded too far, too fast, only to crash amid scandal, had temporarily tarbrushed a fledgling industry, but too much else was happening for it to stay down for long. In 1994, it looked like my home state of Connecticut might become an EV center. Solectria was negotiating for a $1.2 million state grant to build a factory in Waterbury for the Sunrise, its very sleek new battery EV. James Worden announced dramatically that he could sell the lightweight carbon fiber–bodied car for $20,000—if he could get orders for twenty thousand cars. But without an industry partner, the plans for the Sunrise fell by the wayside.

It looked like the Hubbell EV Power Port, made in Bridgeport, could triumph over the Hughes paddle charger I saw demonstrated in Torrance, but that didn't happen either. The state's EV boosters formed the Connecticut Clean Transportation Association and fielded small grants from the state's electric utilities. A Central Connecticut State University professor, Daryll Dowty, designed an electric motorcycle named the Envirocycle and toyed with the idea of manufacturing it. Meanwhile, an East Lyme business was trying to put EV shuttle buses in place at such leading Connecticut tourist attractions as Mystic Seaport and at Foxwoods, the largest-grossing casino in the world.

I rode down the tony streets of Greenwich in the Volkswagen Solar Flair, a Rabbit diesel converted to electricity by students at Greenwich High School. It rattled and could barely climb hills, but it definitely worked. We used the Flair to go visit Vinny DeMarco, a foreign car mechanic who had gotten religion about EVs and was trying to convince the Greenwich Transportation Board to help subsidize an electric commuter service in the town.

In Connecticut, as in California, New York, and other states, entrepreneurs were starting to see a profitable future for EVs. They were even getting the first cars on the market, two years ahead of the first Big Three entries. The wind seemed to be in their sails, because of the California legislation and a growing public interest.

Despite all that, I had gone to California in an optimistic mood and came back vaguely troubled. The Ford Ecostar was clearly superior to the independent entries—a finished, production-standard vehicle with the kind of attention to detail you'd expect from a major manufacturer. Its competitors were hodgepodge conversions. And even if their cars had been better, how would the small companies, lacking any kind of national distribution, get them in front of the buying public? I had the sense the independents would quickly be overtaken by even a halfhearted marketing effort from Ford, GM, and Chrysler. And that, indeed, is what happened. The big companies were about to show off what they'd been working on behind closed doors.

ELECTRIC VEHICLE SHOWS

Until the 1990s, EV showcases were small, insular affairs, occupying the kind of modest hall also rented out to countywide spelling bees. At the major auto shows in Detroit, Frankfurt, Tokyo, and New York, the EVs were off in a corner, presided over by a forlorn company loyalist who was probably longing for another assignment.

That's all changed now. Today's international auto shows put their new green designs front and center. At the 1999 Greater Los Angeles Auto Show, for instance, six of the world's biggest automakers showed off EVs, and a conference called "Automobiles and the Environment" featured a keynote address from Rocky Mountain Institute co-founder Amory Lovins, proponent of the lightweight fuel-cell-powered "hypercar." EVs were enshrined as design milestones at a 1999 Museum of Modern Art show in New York entitled "Different Roads: Automobiles for the Next Century."

As interest in EVs has skyrocketed, trade organizations such as the highly professional Electric Vehicle Association of the Americas (EVAA) have taken what was an isolated industry and unified it through increasingly high-profile annual shows held all over the world, from Brussels to Beijing. Instead of a handful of small exhibits from struggling vendors, there are big displays from mainstream names such as Honda, Ford, GM, and Toyota, free test rides, guest appearances by celebrities, gala banquets, and complimentary mouse pads. In addition, clean-power partisans such as the Northeast Sustainable Energy Association (NESEA) attract throngs to crowd-pleasing events like the annual American Tour de Sol, an EV "race" (in reality, more of an efficiency challenge). Early events attracted entrants from college engineering departments and conversion companies. In 1989, only four challengers competed in the first Tour de Sol; in 1994, it attracted sixty. These days, the independents go up against factory-backed teams from the major automakers. Racers can also sign on for endurance tests such as Phoenix, Arizona's Solar Electric 500.

By 1997, EVAA's Electric Vehicle Shows (EVS) were generating substantial buzz. In December, I went down to Orlando, Florida, for EVS 14, where I'd finally have a chance to drive some of the big companies' EVs, plus Toyota's revolutionary hybrid Prius.

Orlando is hardly a green mecca, though host Disney World did partner with GM to have Mickey's maintenance workers drive electric S-10 pickups. Disney does want to be thought sustainable these days, as it tries to erase unpleasant memories of the company that was fined $550,000 by the Environmental Protection Agency in 1990 for sewage violations and improper toxic-waste storage in its Orlando operations.

Orlando itself certainly needs sustainable transportation. Interstate 4, which runs through the city, is a gridlocked mess, with idling cars pumping pollutants into skies no longer as blue as they were when Gatorland was the city's principal tourist attraction. When Walt Disney first came to town in 1966, he envisioned an innovative experimental community whose air quality was controlled inside a plastic bubble. Instead, his company waited thirty years and built the town of Celebration, where the real world is kept at bay with heavy security, but pollution (as seen most dramatically in Orlando's lakes, which are breeding abnormal alligators) manages to intrude. Since Orlando isn't likely to build the efficient monorail Disney dreamed of any time soon (a plan for a modern elevated rail system gave way to a touristy trolley line), EVs may help the city breathe a little easier.

And EVs are exactly what came to Orlando in December 1997. If you squinted really hard while watching a steady parade of exotic-looking EVs zoom around the forecourt of the Walt Disney World Dolphin Hotel, you might have thought you were in Tomorrowland.

To me, EVS 14 was a great opportunity to actually get behind the wheel of cars I'd only been reading about. I drove a Toyota Prius around the Disney parking lots, and the most remarkable thing about the car was getting used to Japanese-market controls (steering and turn signals on the right). Driving it was a fairly pleasant, if bland, experience—you could be behind the wheel of a Tercel, for all the dramatic effect. Steering effort was light, acceleration leisurely but steady. A Prius and a Chevy Metro with a tailwind could probably keep pace. The anxious company man in the passenger seat (a fixture in all EV test drives) could have relaxed; nothing fell off, nothing struck a jarring note. The Prius will perform differently under different conditions, and in the confines of a Disney parking lot, full of stops and starts, it switched repeatedly between the electric and gas motors. The changes would have been imperceptible if the car's keeper hadn't pointed them out—it felt like an automatic transmission shifting.

There was very little in the test Prius to give away its status as Car of the Future, except for a color TV screen that displayed technical data. These are apparently popular among Japanese consumers, who also love global positioning systems, but Americans may be distracted by the shifting graphics (which amount to status reports on the two-drive systems). Everyone who drove the Prius thought it was ready to meet the American public, with some adjustments for the vastly different road conditions and much cheaper gas.

In all the hoopla it was easy to overlook Toyota's competent, if unspectacular, RAV4 EV, on sale to large fleet buyers in California. The rattle-free car I drove gave me a sense of what Toyota could do with a mass-produced EV, and it was so easy to pilot that *Saturday Night Live*'s Toonces, the driving cat, could get one safely home. After a few miles in it, I felt completely at home.

EVS 14 was marked by new model introductions and an air of celebration. In a 1968 issue of *Motor Trend,* a story headlined "Electric Cars: Anytime You Want Them!" proclaimed, "The scientific state of the art is such that you could own and drive an electric car tomorrow." But after decades of dashed hopes, there was finally a palpable sense that this eternal car of tomorrow had really arrived.

In 1998, the show scene shifted to Europe, where a strong interest in fuel cells and electric "city cars" was growing. It was perhaps fitting that GM chose to introduce its groundbreaking methanol fuel-cell vehicle (a Zafira minivan reduced to a two-seater by all the hardware) at the 1998 Paris Auto Show, since it was jointly developed by U.S. and German GM engineers and was built on the platform of a German Opel.

The theme of EVS 15, held in Brussels, Belgium, was "A Future for the City," and it reflected the many European initiatives, in small towns and big cities like Paris (which has 180 charging stations), to become EV-friendly. You can now rent EVs in half a dozen European cities (and in Los Angeles as well, through Budget Rent-a-Car). Some European town centers are now "ICE-free" (no internal-combustion engines), and hybrid city buses are coming to Athens and Luxembourg.

EVS 15 was a blizzard of paper, as dozens of engineers from all over the world presented technical talks on everything from new battery developments to fuel-cell timetables. The general public had more chance to get involved in the Fourth Annual International Electric Car Rally, held in Monaco right after the Brussels event. Interest in EVs is high in the tiny harborside principality, whose bowl geography dooms it to serious smog problems. Rally attendees got to mix with royalty, since Prince Albert Rainier took a keen interest in the proceedings, as did Germany's Prince Eberhard von Württemberg, who is marketing his own EV. It being Monaco, electric boats were also much in evidence, and ralliers passed around a picture of Microsoft's Bill Gates riding in one. Toyota, which brought a large factory team, won the rally, as it did the two previous years.

While U.S. events are mostly about technology, European ones are more about infrastructure, building an EV-friendly base through local initiatives and public participation. High gasoline prices predispose Europeans to small, fuel-efficient cars, and European cities lack the sprawl that leads to long commutes. EVs are popping up everywhere on the continent, particularly in the form of small delivery vehicles and buses. In the United States, only California can be said to have performed such diligent spade work. Can EVs take off without a nationwide grid of charging stations and service networks? Some argue that with the cars in use, the support network will fall into place around it. But will consumers make such a leap of faith?

At the 1997 Orlando show, Frank Pereira, GM's then brand manager for advanced technology, had compared the EV to the cellular telephone, which, he noted, few people bought in its first years on the market. "Luddites claimed they'd never catch on," he said, adding that personal computers and microwave ovens weren't instant hits either.

According to *The Wall Street Journal,* the slow pace of EV acceptance is nothing unusual. It took the cellular phone thirteen years to reach 25 percent of the country's consumers, the newspaper reported. The personal computer took fifteen years, the TV twenty-six, the radio twenty-two, and the airplane fifty-four.[9] Most of those products, after some initial excitement generated by early adopters, began the slow and steady climb to becoming indispensable household products. The EV, of course, is not a new product, but its appearance as a modern consumable should probably be dated to GM's introduction of the EV1 in 1996. If EVs are a big hit by 2010, then they're probably right on schedule.

The fledgling EV industry is offering something every bit as important as the VCR and the cellular phone—clean cars, at last. And the first mainstream customers are undoubtedly satisfied. There just aren't enough of them. The battery EV is still not good enough, not *transparent* enough, to flood the market. By 1999, it was the humble Prius, a car easily overlooked amid the more flamboyant prototype show cars in Detroit and Los Angeles, that seemed destined to establish a firm beachhead for a promised full-fledged invasion of the hybrids and, later, the fuel cells. These were cars that you could just jump into and go, emitting very little pollution in your wake. It's no insult to say that they were on their way to becoming appliances, just like the pop-up toaster.

CHAPTER 5

U-TURN: THE BIG THREE GET SERIOUS ABOUT GREEN CARS

CLEAN CARS LOOK great on paper, and I was accumulating reams of the stuff. By the late spring of 1998, my office was bulging with voluminous files of magazine and newspaper clippings, as well as a six-foot stack of public relations material. I'd also conducted dozens of telephone interviews. But a lot of what I knew at that time was theoretical and involved projections into the near and distant future. I still had not actually *seen* a fuel cell or had more than a brief ride in a hybrid car. I knew that the big auto companies were increasingly willing to put substantial funding into their hybrid and fuel-cell projects, but I as yet had no real sense if a true automotive "revolution" was going to happen, or if this hopeful prospect would disappear into thin air like so much exhaust smoke. There was certainly precedent for that: The journalists of 1973 thought that EVs were imminent, but the return of cheap oil rejuvenated internal combustion.

Battery-powered EVs, symbolized by the GM EV1, launched with such fanfare in 1996, had also proven to be a disappointment. I had become convinced that EVs would take off, this time, only if they were in fact better by most meaningful measures than the cars people were already driving. That meant they had to accelerate faster, operate more quietly, cruise farther in more comfort, and look better than today's Cadillacs, BMWs, and Hondas. The fascinating thing was that both fuel-cell and hybrid cars promised just that.

Was the auto industry serious about all this? I decided I'd have to hit the road and talk face-to-face with the people who were making this profound change happen. Instead of relying on industry backgrounders, I wanted to visit the shop floors where the prototype cars were actually being built, mostly by hand. Setting up a working schedule took some time, since the auto industry is legendary for the protection it affords its trade secrets (with blurred photos of new models being leaked to the car magazines like CIA satellite data). I eventually convinced the gatekeepers that I was essentially benign and got most of the access I sought, both to prototype cars and key executives. Feigning complete technical ignorance helped.

I decided to start my trip in Canada, home of Ballard Power Systems, the company that was making the fuel cell practical. While Ballard was certainly not the only important fuel-cell player, it was probably the most focused. Its single-minded concentration on consumer-size PEM stacks (hundreds of wafer-thin individual cells bolted together) had paid off with breakthrough after breakthrough, attracting all the major automotive players to its home in a Vancouver suburb. And now it was attracting me, too.

THE BOYS FROM BALLARD

On even the nicest summer days, there's a bit of haze over Vancouver, a city otherwise blessed with great natural beauty, splendid mountains, and picturesque bays. But Vancouver lies in a valley that's like a natural funnel, and the smog that settles there can't clear the North Shore mountains; indeed, it shields those mountains from public view in the warmer summer months.

When I visited Vancouver in the summer of 1998, a heat wave had led to a three-day air-quality advisory, and Canadian environmentalist David Suzuki had held a press conference, with those hazy mountains as a backdrop, calling for new government regulations to reduce fossil-fuel consumption and build more rapid transit.

The environmentalists in Vancouver don't talk much about smokestack pollution because Vancouver has a relatively clean combination of hydroelectric power and natural-gas generation. "The ground-level ozone pollution is mainly from cars and trucks," says Cheeying Ho, executive director of Vancouver's Better Environmentally Sound Transportation (BEST).

Vancouver is lucky in that its electric trolley service is still thriving, and buses connected to overhead wires make up most of the municipal fleet. But the rest of the buses are dirty and aging two-stroke diesels.

The good news is that BC Transit has replaced forty-four of its smoky buses. The replacements are diesels, too, and that's drawn some criticism from environmentalists. But these are low-emission, low-sulfur, four-stroke diesels, emitting a fraction as much hydrocarbon and nitrogen oxide as their predecessors.[1] For her part, Ho isn't too worried that the city's new buses will have exhaust pipes. "The public has a lot of misinformation about diesels. They think they're all dirty," she says.

But in Vancouver, the public hasn't even focused on the new buses, good or bad. Instead, BC Transit is perceived to have rammed its plan through over the objection of protective neighborhood councils that didn't want to be on the local high-speed bus routes at all. It's an instructive case for mass-transit advocates who think getting people out of private cars is easily accomplished. "We have become a lightning rod for resentment," says Stephen Rees, senior transportation planner at BC Transit. "People see this as an opportunity to vent their rage."

All of this was prologue to a visit to Ballard Power Systems, which is in the eastern Vancouver suburb of Burnaby. Ballard also has a project with BC Transit, but it's not likely to result in any public indignation. The company is the world leader in making fuel cells, and it has installed three of them in city buses that, at the time of my visit, were about to take to the streets for the first time. The fuel-cell-powered New Flier buses are much cleaner than the new diesels. In fact, they're not contributing to Vancouver's smog problem at all. Along with a similar test under way in Chicago, where the modified city buses are called Green Machines, the Vancouver pilot program was the world's first acid test of fuel-cell vehicles. The buses, which are practically noiseless except for the whine of their air compressors, have a range of 250 miles, easily trumping the most modern battery-powered electric. As they trundle down Vancouver's crowded streets, the Ballard fuel-cell buses just might be making history.

Ballard doesn't build cars, trucks, or buses: Its sole product is the fuel cell in all its many applications, plus the ancillary equipment to make it work. But because Ballard's technology is so advanced, its fuel-cell sales are enough to make it one of the fastest-growing automotive suppliers in the world. It's not surprising that executives from eight of the nine largest carmakers have racked up frequent-flier miles visiting the place—often coming away with fuel-cell contracts. Not bad for a company founded in 1979 to build rechargeable lithium batteries for smoke detectors.

With 450 employees and research operations in Germany and California as well as several locations in Canada, Ballard today is a beehive of activity. The daily sign-in book is full of Japanese and German names. Fuel-cell people, you soon learn, take a lot of meetings, and interpreters are probably helpful.

My first meeting was with vice president Paul Lancaster, a slight, bearded man with the satisfied look of a professor who'd just been granted tenure. Although Ballard has never actually made a profit (except under Canadian accounting rules), its stock has made meteoric gains, and if Lancaster owns any shares at all he has good reason to smile. "We will soon have a fuel cell ready for volume production, up to two hundred and fifty thousand annually," he said with an expansive wave around the company conference room. "I don't think there's any doubt that fuel cells will come in and change everything. I have a one hundred percent certainty that will happen. The question is, when? Everything is in place. From there, it's dependent on the auto industry. When will they first be brought to market? What kind of infrastructure will they have? Sure it's competitive, but for fuel cells to work, you need a lot of companies making them at the same time. That means we have to be one hundred percent independent. In addition to our Alliance partners DaimlerChrysler and Ford, Honda has been a customer, Nissan, Mazda, Volvo, and Volkswagen. We'll sell our technology to any and all."

Although it's very successful today, Ballard's initial goals were green, not gold. The company's founder, Canadian engineering geologist Geoffrey Ballard, had worked for the U.S. Department of Energy but left after concluding that the government agency wasn't really interested in alternative power. Ballard is a serious environmentalist, and the desire to save the planet was very much part of his thinking when he first began to investigate the fuel cell's potential. "The idea was to change the way the world used power," he said. "Literally, through the internal-combustion engine, 20 percent of the world's population consumes 90 percent of the energy. When India and China decide that they want the same standard of living that we have, and they're making that decision now, they have to burn energy. The standard of living is tied irrevocably to the consumption of energy, so you must have a clean energy source or else, when the rest . . . of the world wants to have what we've got, the world won't be able to sustain it."[2]

In the early days, Ballard was so poor that it maintained offices only through the good graces of its landlord. Ballard himself had to be supported by his wife, Shelagh, who bought a pub to keep money coming in.

The turning point was in 1983, when the struggling Ballard was approached by the Canadian Department of Defense. "They were interested in fuel cells," Lancaster says, "and since they're operationally similar to batteries, they thought we might want to bid on a contract." Ballard admits ruefully that he was initially skeptical. "I kept thinking it would be the battery."

Ironically, it was an American company, General Electric, that had done most of the pioneering work on PEM fuel cells in the early 1960s,[3] but GE wasn't thinking about powering cars or decentralizing home electricity. Then far more of a military contractor than it is today, GE developed its PEM cells for the Gemini space program. But because the cells it built cost hundreds of thousands of dollars for every kilowatt they generated, GE saw no practical consumer application. And it let many of its patents run out. (In a further irony, GE recently announced it was getting back into fuel cells—as a distributor for other companies' products.)

Ballard got very excited about the fuel cell's potential, though ill health forced the company's founder out of operational involvement just as the new technology's potential was being recognized. A pivotal moment was a typically blustery day in January 1993, when Ballard's prototype fuel-cell bus, a running and driving 125-horsepower advertisement for the company's technology, first moved under its own power. That same day, Ballard had been attending an international energy conference, where he was confidently told that it would take years for his company to build a functioning model. When those same transit officials saw Ballard's bus drive by, he says, "their mouths dropped open."

Buses are, obviously, an ideal platform for a hydrogen-powered fuel cell, particularly the bulky technology that existed in 1993, because they offer a huge roof that can be used for fuel storage, a wide, flat floor that's perfect for batteries, and an enormous rear engine compartment that easily adopts to house the cell "engine." What's more, municipal buses are usually run out of a centralized depot that can be used for large-scale hydrogen production and storage. Ballard's bus, like the International Fuel Cells and Daimler-Benz buses that appeared later, proved a notable "proof of concept" for skeptics who thought that drivable prototypes were still years ahead.

The three fuel-cell-powered buses now in service in Chicago, with their roof-mounted silver spouts blowing off the occasional puff of steam, have become local celebrities. Chicago's Mayor Richard Daley quaffed a glass of water produced by a Ballard fuel-cell stack and proclaimed, "Not bad." More important, the buses serve as invaluable rolling test beds for Ballard's engineers to gather operating data. "We've had some bus stoppages, a few due to stack problems," Lancaster admits. "But of forty-five hundred cells in the three buses, we've had trouble with only ten of them. Our main problems have been with the air-conditioning and the air compressor for the brakes—systems we didn't build."

Getting a fuel cell to work on a test bed is far different from having it perform reliably, day after day, in real-life high-stress work conditions. Lancaster says the Chicago drivers want fast acceleration under twenty-five miles per hour, to help them merge into traffic. The mechanics, it seems, are happy if everything's familiar. Apparently uncomfortable with the modern sensors built into the vehicle, the Chicago garages installed old-fashioned gauges. Which leaked.

Ballard's buses are relatively simple in that they run on pure hydrogen, without the need for a reformer, and the fuel cells develop enough power on their own without the need to be supplemented by big, space-grabbing, and weight-adding battery packs. By 1995, the company had developed a fuel-cell stack producing the equivalent of 275 horsepower, using a fraction of the space the 125-horsepower 1993 model had taken up.

Ballard hasn't built a fuel-cell car, and it isn't going to. The company won't become a carmaker, but it will become a manufacturer—of fuel cells themselves. In addition to the stack for transit buses, Ballard is building 50- to 100-kilowatt systems for cars, and a small, under-two-kilowatt portable unit that could power a laptop computer or fit in a soldier's backpack. A separate subsidiary builds stationary 250-kilowatt power plants to run hospitals and factories, and a much smaller 10-kilowatt model that can take the average suburban house off the utility grid.

Ballard builds fuel cells, but its partners in what's called the Alliance—Ford, Volvo, and DaimlerChrysler—will handle every other component of the system, from the car body to the electric motor drive. The cell can be likened to the car's engine block: It needs cooling, control, and fuel processing to perform, so a lot of systems-integration work has to be done. The goal, says Lancaster, is a fuel-cell car that's competitive in price with internal combustion. "Some early adopters will pay a premium for new technology," he says, "but people will not pay double for cars. In the showroom, if fuel cell and internal-combustion cars with the same refueling, power, and convenience sit side by side, and one costs much more than the other, we'll lose out. It has to be a no-brainer for people. But there's really no reason why fuel-cell cars should cost more."

Lancaster seems very confident that Ballard will have a cost-effective fuel-cell prototype in 2000 or soon after, but will there be an infrastructure to support it? The main problem is fuel storage, and Lancaster is skeptical that high-density storage of hydrogen gas will be made practical anytime soon. Liquid hydrogen presents just as many problems, he says. Lancaster thinks that, in what may turn out to be a transitional phase, the first fuel-cell vehicles will run on methanol, which is relatively simple to "reform" and doesn't present too big a change from our current system. Methanol is both toxic and very corrosive, but our current gas stations could be retrofitted to handle it, with some strengthening of the fuel lines. (Many gas station tanks are already methanol-compliant.) Obviously, companies like Ballard are worried that fuel infrastructure problems will strangle their infant industry in its crib.

If Lancaster seemed confident, I was unprepared for the dynamic certainty of Firoz Rasul, the Kenyan-born Indian who is now the company's president and CEO. It was actually Rasul who made the decision to focus the company on fuel cells in 1989, at a time when they were getting very little attention. A tall, handsome man with a bristling white mustache contrasting with his dark skin, Rasul is convinced that fuel cells will become a big energy player, and soon. "By 2030, I think that one in three or even one in two of the cars on the road will be fuel-cell vehicles. There are obstacles and challenges to that happening, but I don't see anything insurmountable. We have to reduce the cost of our materials, and develop high-volume manufacturing processes. We've already demonstrated that the performance is achievable."

Like Lancaster, Rasul thinks the initial infrastructure for fuel cells will be based on methanol, though gasoline remains at least a possibility. "We're rooting both of them on," he says. "If gasoline is the fuel, great, because it will be hard to do if we don't have the oil companies on board. The fuel is still a problem. We have to learn how to work with a gaseous fuel that's lighter than air, burns quickly without a flame, doesn't puddle on the ground, and dissipates quickly; but we're confident fuel cells can be launched without that problem solved."

By 2003, Rasul thinks, the world's industries will have made between 100 and 150 experimental prototype fuel-cell cars, each hand built and each representing a substantial investment. Ballard's buses, for instance, are $1 million each, which helps to explain why only six were on the road in 1998. The progress of the fuel-cell car will be slow and steady, he thinks. Ballard is looking beyond the Chicago and Vancouver demonstration projects to an ambitious project called the California Fuel Cell Partnership, which will put fifty fuel-cell test vehicles—five cars and twenty-five buses—on California's roads between 2001 and 2003.

Other commercial applications could come sooner. The portable fuel cell, ready to keep a laptop going for twenty hours, will be marketable in 2001, the same year the first commercial buses are expected to appear. And Rasul is equally convinced that a large stationary power market will emerge soon.

"Stationary power is so huge," Rasul says. "We're talking about the whole Third World. In countries where the cost of putting a grid in place is out of reach, they can leapfrog the wires and go straight to wireless technology. Most countries in Southeast Asia and South America have huge supplies of natural gas, which is ideal for fuel cells. Other countries will take advantage of available biomass and methane, even attach our cells to solid-waste-treatment plants. One of the most wonderful things about fuel cells is that they can run on anything."

Rasul doesn't see Ballard as a bit player. "Our vision is to create the next Intel and provide this technology to the whole world," he says. "The fuel cell has as many applications as your mind can imagine. Think about when the microprocessor was invented. The fuel cell will transform the world in many ways, turn a lot of things on their heads."

When Ballard first brought its fuel-cell business plan to venture capitalists, they saw these chemical engines as restricted to a niche market. "We said, 'No, it's a much bigger opportunity,' " Rasul says. "We told them we wanted to be a manufacturer, not just a company that licensed its technology. If we stopped actually making fuel cells, we would stop being a leader in the field." Ballard's first production lines are in the Vancouver area; the small German operation is limited to research. But the company has ambitious expansion plans that see it owning manufacturing operations around the world, with a variety of plants close to its auto-industry customers.

Lancaster and Rasul gave me a tour of the Burnaby facility, which houses research and manufacturing units all together in one enormous open room, like an aircraft hangar. In a scene right out of the movie *Brazil,* the high ceiling was snaked with wires and pipes, delivering electricity, hydrogen, nitrogen, and helium to the testing stations. The first thing the boys from Ballard showed me was a blue box the size and shape of a refrigerator. It was a prototype of the 10-kilowatt fuel cell that Ballard would like to see in every home. Turned on, the box was practically noiseless, except for an occasional whoosh of its built-in compressor expelling excess hydrogen and air. The machine was hooked up to illuminate a bank of spotlights, making a dramatic display, but it could just as easily have been powering your teenager's hair dryer. Ballard will probably have a lot of competition making units like this one, because it promises to deliver cheap and exceptionally clean electric power, running on the natural gas lines that are already in most American homes.

The next stop was a table showcasing an array of fuel-cell stacks. Here was a model of the two-kilowatt portable unit. At the time of my visit, the pack, about the size of a large flashlight, was disappointingly heavy, far outweighing the laptop computer it would presumably power. Future models would be much lighter, the executives said.

Such progressive miniaturization characterized the successive generations of automotive stacks on the table. Though each was approximately the same size, the amount of power they produced increased exponentially. A 1990 stack, about the size of a small air conditioner on its side, generated just three kilowatts, enough power to run some household appliances. By 1993, a similar stack was producing 10 kilowatts. Just two years later, a 150-pound stack could push out 25 kilowatts, and two of them were small enough to fit inside of and power a compact car like the Mercedes-Benz A-Class. But the most recent 60-kilowatt Ballard design, no bigger, could boot your fuel-cell car into the passing lane.

In a corner, Ballard mechanics were working on a pair of fuel-cell buses. One, in the red, white, and blue Vancouver municipal colors, was stripped down to expose an array of twenty 10-kilowatt stacks. In this configuration, the bus had a range of 250 miles and a top speed of sixty-three, certainly respectable for city duty.

Ballard spokeswoman Debbie Roman (whose father, ironically, is a former Houston-based oil executive) took me out to a municipal transit center in Port Coquitlam, another Vancouver suburb, where drivers were being trained to operate the buses, which now make routine runs down Broadway. The center looked like any other bus garage, complete with a rowdy workers' cafeteria, but not many bus depots have large hydrogen-generating stations.

Our driver, BC Transit operation supervisor Jim Kelly, knew what he was doing—theoretically—but he had never actually piloted a fuel-cell bus before. Sitting among long rows of diesel craft, the sixty-person capacity bus didn't stand out, but ample graphics left no doubt about its provenance. It felt momentous to sit there next to Kelly, a veteran of fourteen years behind the wheel, as he tentatively pushed a few buttons, attempting to bring the beast to life. Unfortunately, despite Kelly's best efforts, the electrons weren't flowing. Kelly and Roman huddled, then made a fairly frantic phone call back to Burnaby. It turns out that they'd forgotten to throw a knife switch that provided security for the batteries. Oh. With that remedied, the bus roared to life. Actually it hissed to life, the sound coming from the compressor that gradually pumped up the air suspension.

"It's just like driving a regular bus," Kelly said as we scooted around the parking lot. "You forget it's unique." Unlike many electric cars, the bus even offered that familiar internal-combustion phenomenon known as "creep"—let your foot off the "gas" pedal in drive and it inches forward. Having driven a school bus in my youth, I was tempted to ask for a turn at the wheel, but before I got a chance the bus suddenly sputtered to a halt. "The sensors are ultrasensitive and designed to shut the system down if they detect any abnormality," Roman explained apologetically. "It's usually nothing serious." Something similar happened to poor Ballard in 1993, during a distinguished unveiling ceremony at Vancouver's Science World.[4] Five minutes before Ballard's bus was to drive up on a platform occupied by the then premier of Canada, an incorrectly sized bolt failed and cut power. Without anyone in the crowd noticing, Ballard's intrepid engineers pushed the bus onto the platform.

I came away from my visit to Ballard exhilarated by the experience of a small but innovative company pushing the envelope on new technology, while retaining the environmental mission that led to its birth. While independent car companies were proving themselves unequal to the task of leading the charge on EVs, Ballard was unquestionably in the lead on fuel cells, forcing the auto industry to join on as partners or get left in the company's very clean exhaust.

PLUG POWER'S GRAND DESIGNS

Around the time I went to Ballard, in the early summer of 1998, I read a *New York Times* story[5] about Plug Power, a small company in Latham, New York, near Schenectady, that had made a significant breakthrough by powering a house with a fuel cell. I later learned that, despite the claims of company press releases, this was not quite unprecedented—both Ballard and International Fuel Cells claim to have pulled off the same trick decades earlier. Since Plug Power had also been developing fuel cells for cars, I drove up to Latham for a meeting with president and CEO Gary Mittleman, a lanky six-footer with the unruly gray hair of the renegade inventor.

A graduate mechanical engineer, Mittleman has had a varied business career that's taken him from Ameritech, which he got into the credit card business, to American Can, where he helped sell off the Dixie Cup division. He founded Plug Power in 1997 after working for its parent company, DTE Energy. "The auto market is great, and it remains so," Mittleman says. "All kinds of companies and government agencies are very interested. But the automotive technology is very challenging."

Automotive fuel cells, he noted, have to be both small and lightweight; they have to be shock-resistant and work in subfreezing temperatures, as well as on 100-degree days. And the prices have to be very low, as low as $50 a kilowatt, the equivalent cost of running an internal-combustion car.

Like Ballard, Plug Power has no intentions of building a fuel-cell car. Instead, its plans are based on working with a major automotive partner. Through the federal Partnership for a New Generation of Vehicles, it has a loose alliance with Ford, and it demonstrated an advanced automotive fuel cell to that company in early 1999. Plug Power's car unit attains 75 percent of its power in less than one-tenth of a second, a major breakthrough for slow-to-start fuel cells.

Ford was undoubtedly impressed, but the company has many alliances, and many suppliers. The car business, Plug Power found out, is high-risk, high-reward. With the right automotive partner, Plug Power will become a big player. Without it, the company will be an also-ran. But in the meantime, it's growing rapidly, from 22 employees at its founding to 150 a year later, aided by a $15 million federal grant and a new alliance with General Electric.

In 1997, Plug Power was on the team—with Boston's Arthur D. Little and the Department of Energy—that first successfully extracted hydrogen from gasoline, a dramatic breakthrough. Mittleman thinks that fuel-cell cars could be in production by 2006, maybe earlier with some strategic government support. "By 2020, I think that half the cars made will have fuel cells, but it may be fifteen to twenty years after that before we've fully converted to hydrogen. Internal-combustion cars are very sophisticated: We understand them; we grew up with them. It's a slow process. Cell phones are still in a growth phase, fifteen years after they were introduced. But by 2020 we'll have had ten full years of intensive fuel-cell development."

I mentioned to Mittleman that futurist Peter Schwartz had predicted "the death of the internal-combustion engine" in favor of fuel cells by 2020. "If pollution and global warming cause governments to take action, he may be right on the money," Mittleman said.

Mittleman thinks we'll see fuel cells in homes, a phenomenon known as "distributed power," much sooner. "In a house, you don't have the issues, like size, weight, and shock resistance, you have with cars," he said. "And it doesn't have to be $50 to $100 a kilowatt. At $1,000 it looks viable, and at $500 it's a home run." Of one hundred million homes in the U.S., he said, seventy-five million have natural gas passing by, giving them a built-in stationary power infrastructure. And like Rasul, he also wants to target countries like India and China, which are bounding ahead in power demand, without large national grids. Plug Power thinks it can make home fuel cells viable by 2002, at a cost of no more than $5,000 each.

Mittleman was the first, but not the last, fuel-cell executive I met who was reluctant to give me a tour for fear of revealing his proprietary secrets. He needn't have worried, since I'm hardly technical enough to notice tiny incremental changes in his designs. After some internal debate, he assented and led me through the usual maze of offices and test stations. The 55,000-square-foot space housed men in lab coats, lots of Rube Goldberg wiring, shiny metal gas pipes, and a machine shop to fabricate parts. I was quickly marched past a very clean room where automotive fuel-cell development went on.

Plug Power's demonstration house, a brick-faced ranch, is walking distance from company headquarters. The seven-kilowatt Plug Power 7000 fuel cell, about the size of a large copy machine, sat in the breezeway. Plug Power's Richard Maddaloni—young, Italian, very serious, and decked out in a Polo shirt—presided over a small repair. He was full of confidence about the future of fuel cells. A mechanical engineer like Mittleman, he's hoping to make fuel cells his career. "It's a growing industry," he said. "I come from doing vibration testing on rotating turbines, and this is certainly more interesting."

Despite his stacks' temporary performance problems, Maddaloni is a true believer. "Fuel cells will soon be part of our everyday lives," he said. "The technology is certainly here today, but what happens depends on how well we communicate the story."

The argument that a growing market for home-power fuel cells will accelerate automotive development is widely circulated but, despite all the difficulties, there's a stronger sense of urgency in the car industry. State and federal clean-air mandates are in the driver's seat for the fuel-cell car, but no such demanding deadline looms over emissions from the home furnace. No doubt the two industries are complementary.

Mass production in any form would certainly bring the cost of fuel-cell power down dramatically. Mittleman thinks we'll soon be buying fuel cells for many different uses at Wal-Mart. He wants Plug Power to be the first company to sell one million fuel cells, and he's aggressive enough to make that happen. As one of the top five independent companies working on fuel cells for cars, Plug Power is directly and keenly competitive with Ballard, which has had more luck in securing auto industry contracts. "Ballard is very strong, and they'll be a winner," Mittleman said. "But we want to be a winner, too."

Neither company can "win" this high-stakes race without auto industry support. The Big Three companies often operate as vertically integrated closed loops, performing all their research and development in-house, but on the fuel-cell project they've put proprietary concerns aside and shown a remarkable willingness to collaborate—with other automakers and with the independents. For once, Detroit is admitting it doesn't have all the answers.

DETROIT: GEARING UP FOR CLEAN CARS

The independent companies, whatever their strengths, aren't *carmakers.* With time and a little luck, they'll come up with a practical, ready-to-manufacture fuel cell. But integrating that cell into a fully functional, reliable automobile that will start on the coldest days is where Detroit comes in.

As an auto columnist, I've made frequent stops in Detroit over the years to attend new car introductions and press events. I remember the stark contrast, when Chrysler was still downtown in Highland Park (it has since fled to upscale Auburn Hills), of driving the then-new Chrysler-Maserati convertible through neighborhoods that remained boarded up from the 1967 riots.

More recently, when the city of Detroit was engaged in an uphill battle to encourage tourism, I was a guest of its Convention and Visitors Bureau. We stayed at the near-empty Renaissance Center, soon to become General Motors world headquarters, and toured city landmarks that all had something to do with the auto industry. Even at the overwhelming Detroit Institute of Arts, the fifth-largest fine arts museum in the world with its Picassos, Van Goghs, and Diego Rivera murals, it was impossible not to think of the incredible auto fortunes that had endowed the place. Edsel Ford's magnificent house in Grosse Pointe was on the tour, as was the Henry Ford Museum, where a number of early electric cars are on display. Magnificent brownstones, some of them now burned out, give a sense of the vibrant neighborhoods that once existed in the city, courtesy of good-paying union jobs and layers of middle management.

Detroit is the hub of a wheel built on the banks of the Detroit River facing Canada, and its suburbs radiate out from it on a series of ring roads. But those suburbs, many of them very affluent (the median family income in Bloomfield, for instance, is $150,000), are largely self-contained, with their own museums, hospitals, and parks. Some suburbanites boast that it's been years since they were within the city limits. That attitude is reflected in the auto industry, which is of Detroit but no longer really in it.

GM: THE GIANT STIRS

While the auto industry has largely turned its back on the city that gave it birth, GM has taken a stand by staying. Ron Williams, the former owner of Detroit's alternative *Metro Times* and a fierce city partisan who located his own newsroom downtown, gives the auto giant a lot of credit. The company had planned to build a new headquarters out in the greensward, but instead bought (for $72 million) and transformed into executive offices the troubled Renaissance hotel, mall, and convention center complex, which had cost $350 million to build in the late 1970s. "It was extremely good news for Detroit," Williams says. "It's driven real estate prices up in certain neighborhoods." Two new city casinos probably haven't hurt, either.

I arrived in Detroit in August of 1998, just as a massive GM strike had ended, costing the company $2 billion and prompting a lot of hand-wringing about where the company was going. "GM: It'll Be a Long Road Back" proclaimed *Business Week*.[6] The newspapers were full of the news that GM had spun off its $31 billion Delphi parts unit (which, in an unusual display of cross-industry cooperation, supplied components to Chrysler's EVs), and Chairman Jack Smith had just announced that the company would build new U.S. factories. That was good news for American operations, but it still left GM's Advanced Technology Vehicles (ATV) division, which produces its EV prototypes, including hybrid cars and fuel cells, a little unsettled. Many at ATV consider Jack Smith, who is on record as predicting "a slow phase-off of the internal-combustion engine," to be their protector and advocate, and the strike may have cost him political support within the company. If he was ousted, they worried, they would be too. Having spent $1 billion since 1990 on the EV1 and related products, ATV certainly wasn't a sideline, but its hold on the attention of the world's largest industrial corporation depends on top-level support.

Detroit's EV divisions are funded by a fear that history will repeat itself. The ground can shift rapidly in the automotive industry, which is dependent on long lead times for new models. In 1975, when memories of the oil embargo were still fresh but Detroit's cars were bigger than ever, there was a record sales year for the Volkswagen Beetle, as well as new inroads by upstarts Toyota and Honda, whose sales grew to 100,000 that year. GM's profits dropped 35 percent, forcing the company to temporarily close fifteen of its twenty-two assembly plants.

Caught with nothing but "full-size" cars in its inventory, GM launched a crash course to build an economy model, surfacing with the less-than-inspired Chevette, which had the distinction of being the smallest Chevy ever. Built on the platform of Germany's Opel Kadett, the four-cylinder, 52-horsepower Chevette managed a respectable thirty-five miles per gallon when it appeared in the Bicentennial Year. In just 1976, 188,000 of the little hatchbacks were sold, at least putting GM back in the game.

For GM, the timing probably seemed right for EVs. In the absence of Big Three competition, tiny start-up companies had appeared, offering pasted-together conversions and cute electric commuter vehicles like the CitiCar, produced by Sebring-Vanguard, which for a shining moment was the fifth largest automaker in the United States.

GM decided to build an EV of its own, the Electrovette, a logical extension of the Chevette program. GM announced that it would build the car in 1980, making a return to plug-in power after seventy years away. In place of the Chevette's anemic four-cylinder engine was a tiny DC electric motor powered by novel zinc nickel oxide batteries. To head the program, GM chose a dedicated engineer named Ken Baker, who, ironically, would survive the debacle and go on to head the successful EV1 program.[7]

The Electrovette wasn't so lucky. In those days before auto electronics, it was dependent on an unreliable mechanical controller. The batteries were expensive and, though much better than the equivalent lead-acid power pack, still couldn't solve the range problem. Ken Baker's team could never get the Electrovette to run properly, and the project petered out, souring everyone associated with it.

After its launch of the EV1 in 1996 and the debut of a spectacular range of alternative-fueled cars at the 1998 Detroit Auto Show, the company was deeply committed to EVs. Like so many of Detroit's EV incubators, ATV wasn't in Detroit, or in fact anywhere near company headquarters. It was instead tucked away near a gigantic mall in suburban Troy.

I couldn't miss the symbolism as I maneuvered the rental car through a series of parking lots overflowing with fifteen-miles-per-gallon sport-utility vehicles. Malls seem to attract these big rigs like bugs to a light. ATV's complex was in an anonymous office building. While there were no big signs on the door, it was easy to tell I was in the right place: not only was I surrounded by GM vehicles, but a special corner of the lot, complete with charging stations, was reserved for EV1s, more than I'd ever seen together before.

Photographs of EV1s in motion, piloted by happy customers, lined the walls. I was taken in to meet Robert Purcell Jr., ATV's executive director, a balding, cherubic fellow with a one-track career in GM's management proving grounds. Like his predecessor, Ken Baker, Purcell is impressively informed about the big picture of EV development and the likely fate of the cars it builds.

"The investment in the EV1 gave us the hyperefficient architecture for a whole range of EVs," he said. "We were able to reduce energy consumption, mass, and accessory loads, and improve aerodynamics, rolling resistance, and driveline efficiency. The structure is an aluminum space frame, and the body, without batteries, weighs only eighteen hundred pounds."

I noted Purcell's use of the phrase "hyper" and asked him if, in fact, GM wanted to build the fuel-cell-powered "hypercar" designed by Rocky Mountain Institute guru Amory Lovins, the visionary whose single-minded advocacy of ultralight fuel-cell vehicles has been credited with sparking the current generation of clean cars. Lovins had told me he's "one hundred percent certain" that we'll eventually have a hydrogen-based economy. "I've worked with Lovins closely, but I'm not one hundred percent certain the sun will come up," Purcell said. "If we look at the ideal scenario for personal transportation, hydrogen has a lot of advantages. But the problem is the hydrogen infrastructure. Hydrogen is difficult to handle; it is, for instance, flammable at just four percent concentrations. There are unique challenges. Over a thirty- to fifty-year horizon we'll quite possibly resolve those challenges."

Looking ahead, Purcell sees "a decreased reliance on fossil fuels, with electrics, hybrids, and a much-improved internal-combustion engine. There's no way internal combustion will remain dominant, but it's not clear what technology will replace it. Our mission is to create options." Actual production of fuel-cell cars, he thinks, is eight or nine years away. "We talk about being production-ready by 2003, but that's demonstration-level hardware. If Ballard can deliver a cost-competitive fuel-cell stack before then, they will be in a very competitive position."

Purcell agrees with me that the present period can be likened to 1900, when gasoline, electric, and steam cars all vied for market share, with the public—and the industry—undecided about the future. He also agrees that, in the great sweep of history, our romance with fossil fuels could turn out to be incredibly brief, not much longer than a single century. "Fifty years from now we'll look back and see the fossil-fuel era as this much," he said, holding up two closely spaced fingers. "But thinking in too long a framework is also a handicap. In the really long term, we're all dead."

With some justification, Purcell points out that by being first to the market with the EV1, GM made a lot of things happen. "The 1998 California zero-emission mandate was a direct result of GM saying at the Los Angeles Auto Show in 1990 that we would have an electric car on the market by the mid-1990s. The legislators took it as a sign that the auto industry was willing to cooperate, and CARB announced its mandate a few months later, in September of 1990. In 1996, when we rolled the EV1 out, I guarantee that it accelerated the pace of the hybrid Toyota Prius. And now every auto show is full of green cars."

ATV's Dick Thompson took me on a tour of the facility, which overflowed with young engineers in jeans. In one large room, I saw the displays from the 1998 Detroit Auto Show, wire-frame "living vehicle" versions of what could eventually be production hybrid and fuel-cell cars. The bronze parallel hybrid car, a stretched EV1 that was driven triumphantly onto the Detroit stage, had now come to rest in pieces in the shop, where also resided a natural-gas-powered Chevy Cavalier and an electric postal van in full federal regalia. I was also shown the latest 50-kilowatt fast charger, which can get an EV1 back on the road in ten minutes.

For a few minutes, I was able to talk with Dr. J. Byron McCormick, the peripatetic codirector of GM's international

Global Alternative Propulsion Center, charged with developing fuel cells for the world market. The center has several offices in the United States and a base of operations in Germany, where, in conjunction with German subsidiary Opel, it built the Zafira fuel-cell minivan. Dr. McCormick came to GM after ten years of development work on PEM fuel cells at Los Alamos National Labs, where many of the major breakthroughs occurred. He was a booster then, and he's a booster now, albeit a cautious one. "We've moved past the 'Can we stick it in a car and drive it down the road?' kind of thing," he says. "For some time now, that hasn't been in doubt. We have running systems now. But if we're going to put these cars into the marketplace, they have to be able to start in the middle of a North Dakota winter. We don't want to generate expectations that are beyond our immediate ability to deliver."

Dr. McCormick is more forthcoming than most at GM about the company's environmental motivations. "There are two underlying concerns behind our work," he says. "One is that world population is continuing to grow rapidly, and the other is that the emerging economies are starting to move into private cars. Both these things are creating more fuel demands and more urgency on the environmental front."

Seeing prototypes and talking policy was one thing, but a trip to nearby Lansing offered a chance to watch history being made, in the form of the first production EV assembly line in seventy years. I arrived on what was, in fact, the second day of 1999 EV1 production. EV1s are made in what used to be the Buick Reatta plant, adjoining the much larger (and more automated) facility where GM assembles Chevrolet Cavalier and Pontiac Sunfire convertibles. These lines had been stilled by the strike only the week before, but now they were humming.

Dave Harper, a dark-haired, no-nonsense career GM executive, manages the plant, and he showed me the highlights of a production process that is like no other in the company. Because only six hundred cars have so far been leased, there's no need for speed: the thirty-three employees virtually hand-build the cars, pushing them down the line and trading off jobs to keep from being bored. It certainly seemed to be a happy group of workers. "We let them take the cars home," Harper said. "They love driving them."

Nearly every aspect of the EV1's construction diverges from standard practice. The steering wheel and seat frames are made of low-weight magnesium; the radio antenna is embedded in the roof to reduce drag; the tires are low-rolling-resistance, self-sealing (meaning no spare) Michelins. The car's aerodynamic shape is more slippery than a Corvette's, and it sits only five inches off the ground. There are two thousand spot welds holding the car's aluminum body together, plus a uniquely sticky glue made by Ciba-Geigy that holds the roof in place. "If you get it on your clothes, it won't come off," said Harper.

There is a lot of wiring in the EV1 to run its complicated software and diagnostic equipment, and it looks like a hydra-headed snake going in. The cars come loaded, with air-conditioning and CD players standard, so there's not much for a customer to do but choose the color: red, blue, or green.

A dramatic moment on the line is when the thirteen-hundred-pound battery pack mates with the body, creating an EV that can then be driven to the next station. Without this pack, the car would weigh only fifteen hundred pounds, a state-of-the-art "hypercar" by anyone's reckoning. Getting those batteries to produce more power, weigh less, and take up less space is obviously the biggest challenge with cars like this.

I had driven an EV1 before, but only on a short route that gave me no chance to really open it up. I'd heard that the EV1 was fast, reaching sixty miles per hour in 8.5 seconds (sports car territory), but I hadn't experienced it for myself. With Dave

Harper as my passenger, I stomped the accelerator on a plant-access road. A low, quite pleasant whine welled up from its electric motor, and the car took off. It really did feel like one of the fastest cars I've ever driven, and it was easy to see how lessee Marvin Rush could "blow off" Ferraris. I could tell Harper was getting a bit nervous about my driving. "The cops wait on this road," he told me. I slowed down, then got on the highway for a few exits. After a few miles, I felt completely relaxed. The best electric cars won't startle you with how different they are.

EV1s have an LED readout indicating miles to go before the car is "empty," and when I drove hard the range fell quickly. I was very impressed with the EV1 I drove, but at the same time felt it couldn't escape the inherent range limitations of lead-acid battery-powered cars. Better things, however, were ahead.

ENERGY CONVERSION'S ADVANCED BATTERIES

To more fully understand current thinking about batteries, I drove just down the road from ATV to the fake-wood-paneled offices of Energy Conversion Devices, the tiny company started in 1960 by maverick battery designer Stan Ovshinsky. A smiling, white-maned septuagenarian who exudes a constant sense of wonder, Ovshinsky holds hundreds of patents on his inventions, but it's the nickel metal hydride battery that led, in 1994, to his joining forces with General Motors in a spinoff company called GM-Ovonic. In late 1998, Ovonic battery packs started going into EV1s, giving them a range of up to 160 miles.

Ovshinsky's unlikely partner in this venture is Bob Stempel, who was the chairman of GM until his ouster in 1992 and a critical force in making the EV1 happen.[8] His move from the pinnacle of established power to a risky alternative-energy venture may have surprised many, but it mirrors the postretirement steps taken by Stempel's sometime business partner Lee Iacocca. The mild-mannered Stempel, Ovshinsky, and Ovshinsky's wife, Iris, are a team frequently encountered at energy conferences and EV unveilings. At lunch in their Troy conference room, which, like most rooms in the building, is illustrated with a periodic chart of the chemical elements, the trio talked batteries and the future of the car industry.

Stempel spoke first, in the measured cadences of a practiced public speaker. "We've now had the automobile for one hundred years," he said. "The early days were a competition between gas, steam, and oil, and now we're looking at a competition again, made possible by advanced batteries and lightweight components. Today's gas car is very clean, it's darned good, but it has been developed about as far as it can go. Stan and I are convinced that EVs are now in an important race to the market, and we're eager to see the technology."

Ovshinsky sees the fight for the world's remaining deposits of oil as a constant international flashpoint. "Look at the Caspian Sea, the Middle East, Asia, they've become explosive areas for conflict. We can't afford to keep burning oil, and an alternative is absolutely necessary; otherwise we're just relying on short-range plans and rearguard actions."

Stempel isn't convinced that the world is running out of oil, but he agrees that it's an unsettling force on the world stage. "The peak of the oil curve keeps changing," he said. "They keep finding new dinosaurs, and I think we'll be able to meet our petroleum needs for a long time. But they fight wars over that stuff."

Stempel worked with Ken Baker (who has since joined Energy Conversion himself) on the star-crossed Electrovette and said he "concluded then that there was no future, as long as we had to work with such poor batteries." What changed Stempel's mind was the teardrop-shaped Sunraycer, a 1987 experimental "solar" competition car that GM used to win an environmental race in Australia. "It proved that with solid-state electronics, EVs could be made to work." Now Stempel is back working with GM, and instead of worrying about the bottom line, he's trying to force that directional change in the company's corporate thinking he'd always found so difficult when he was in-house.

Ovshinsky is the first to admit that "the ground is littered" with battery failures. "The carmakers keep saying, 'When will we have a battery?' But the best-kept secret is that we have the range and the distance now." Stempel added, "Lead-acid is all we had for the launch of the EV1, so we went with that, but when we put a nickel metal hydride pack in the car, it went two hundred and one miles on a charge. Everyone agrees that EVs won't work with inferior batteries, but these are superior batteries."

Ovonic batteries were undoubtedly better than the competition, but were they good enough? A major drawback remained cost. Ovshinsky had said he could cut the cost of EV-ready packs from $20,000 to $5,000, but that kind of cost reduction takes much bigger volumes than were being produced. And would EV buyers be satisfied with a range of 160 miles, when their other car can go 300 or 400? Judging by EV1 lease numbers, the answer was no. Most carmakers were starting to look beyond battery EVs, and so was Energy Conversion Devices itself. In the summer of 1999, Stempel announced that the company was working with Royal Dutch Shell on hydrogen stations for fuel-cell cars.

FORD'S BETTER IDEA

In the early 1930s, Henry Ford walked into his company's research lab with a bag of chicken bones, dumped them on a body-shop fabricator's desk, and proclaimed, "See what you can do with these."[9] He later urged his staff to try out cantaloupes, carrots, cornstalks, cabbages, and onions in his search for material with which to build an organic car body.

Ford didn't give up, and he eventually hit upon his dream material: soybean stalks. In 1940, Ford scientists discovered that soybean oil could be used to make a high-quality paint enamel, and also could be molded into a fiber-based plastic. The company proclaimed the material had ten times the shock resistance of steel, and Ford himself delighted in demonstrating that strength by pounding on a soybean trunk lid with an ax. When fourteen of the resulting panels were hung on a metal-framed 1941 Ford, the resulting car weighed only two thousand pounds.[10] We might be driving soybean Fords today, if not for the fact that the new material was found to need a long time to cure, and it did not mold well.

Ford's fascination with electric cars has already been discussed in chapter 1. Unfortunately, the reputation for innovation that pushed Ford to the peak of industrial production in its early years didn't survive its messianic founder, and the company slumbered through the 1940s and 1950s. Even such innovative and trendy cars as the 1964 Mustang were technically rather pedestrian.

But now Ford is changing. Under the internationally minded Scotsman Alex Trotman, whom *Fortune* called the company's "most visionary leader since Henry I,"[11] Ford launched the 2000 program in 1995, designed to unseat archrival GM and close down regional fiefdoms in favor of its long-sought "world car." Ford, not an automotive innovator since the golden days of the Model A, was now pouring money and resources into its new Advanced Vehicle Technology division. And interesting things were coming out. How about a bi-fuel 590-horsepower Super Stallion Mustang that can go from zero to sixty in 3.85 seconds—on ethanol!

In late 1997, Ford announced that it would invest $420 million in a global alliance with what was then Daimler-Benz and Ballard Power Systems. The deal put tiny Ballard on the map, and the company got a vital infusion of capital. (As part of a series of investments, Ford now owns 15.1 percent of Ballard, and DaimlerChrysler 20 percent.)

It was a dramatic development for fuel cells. Jason Mark, a transportation analyst at the Union of Concerned Scientists, proclaimed, "This is real progress. A nearly half-billion-dollar investment is nothing to sneeze at."[12] The investment was nearer $1 billion, considering the $450 million brought to the table by the then Daimler-Benz, and it catapulted the Alliance (Ford, Volvo, and DaimlerChrysler) to the forefront of fuel-cell innovation.

The spokesman for Ford's electric-vehicle program is John Wallace, a tall, thin, charismatic man with a substantial mustache and electric blue eyes to match his cars. Wallace, who has a background as a computer engineer, was the force behind Ford's Ecostar van program, launched in 1993 in response to the announcement of GM's EV1, and most of its green initiatives since. He sits on the board of the Alliance, where his role seems to be making sure that the engineers stay on track to bring a fuel-cell car to the production stage. "I hope we are past the point of any questions about our commitment to the electric vehicle," Wallace said in a 1995 speech.

I met Wallace at the Electric Vehicle Center in Dearborn, literally in the shadow of the company's world headquarters and not far from where Henry Ford I had wielded his ax. Wallace strolled in from a meeting and got right to the point. "Yes, Ford has fuel-cell prototypes right now, and we'll show them when they make good public relations impact. But I'm not interested in nondrivable prototypes—I need real road-ready vehicles. I don't work for the research department, I work for product development, so I don't get to rest until I deliver a car ready to go."

Wallace was bullish on Ballard, praising it, as others have, for its singular focus on PEM cells. "If they meet their target for the next two years," he said, "and the other sectors meet their targets too, well then, this sucker is going out the door."

Wallace had, of course, been to Germany and seen what was then the latest generation of Daimler-Benz's fuel-cell prototypes, NECAR III. This news-making vehicle was a tiny Mercedes-Benz A-Class car stuffed full of Ballard's methanol-reformed fuel-cell system. "But I'm more interested in NECAR VI than I am in NECAR III," Wallace said. "That's the one that might go into production."

And Ford wants to go into production itself, with a fuel-cell family car based on the aluminum-and-composite P2000, which is based on the current Contour but weighs a thousand pounds less. In 1997, it announced that its fuel-cell car would carry compressed hydrogen,[13] but the fuel question is still very much in play, and Wallace says Ford is looking at a number of options.

A key to what Ford might do is provided by Sandy Thomas, vice president of the Virginia-based Directed Technologies, which has done considerable consultant work for the auto giant. Thomas believes strongly that cars can carry hydrogen gas, eliminating the need for costly and bulky reformers. As the co-author of a 1998 technical paper published in the *International Journal of Hydrogen Energy,* he concluded that "the two potential barriers to a viable direct hydrogen fuel-cell vehicle market—onboard hydrogen storage and the hydrogen infrastructure—could be overcome."

In an interview, Thomas said that many studies point to the superiority of direct hydrogen, but this option is still regarded by the automobile industry as less attractive than liquid fuels such as methanol. "You could argue that methanol is the worst of both worlds," Thomas said. "There has to be an onboard reformer, and you have to build a new infrastructure. But methanol does have advantages. There's excess generating capacity, and it's the least expensive to transport."

Thomas conjured up a truly spectacular zero-emissions system of "solar hydrogen" in which the gas fuel is produced from a combination of photovoltaic thermal collectors, wind generators, and biomass. "Imagine," he said, "a motor vehicle fuel so clean-burning that you could drink the effluent from the tailpipe, with urban smog a distant memory."[14] The auto companies have a long way to go before such ideal scenarios can be realized.

In common with GM, which uses the EV1 as the base for its next generation of hybrid and fuel-cell cars, Ford is focusing on the P2000, a sensible-looking but revolutionary lightweight experimental sedan. Equipped with Ford's aluminum 1.2-liter direct-injection DIATA engine, the car gets sixty-three miles per gallon, and it tops that in the hybrid configuration the company prepared simultaneously.

Ford is experimenting with a car powered by direct hydrogen, but early tests showed it to be far from production-ready. Even equipped with a high-compression 5,000-psi tank, the P2000 could only manage a range of fifty miles, although acceleration was reportedly impressive. The engineers' response was to shed even more weight off the car, but they'll try many other things as well.

I had a chance to drive the P2000 as a work-in-progress on a test track near Ford's Dearborn world headquarters in the summer of 1999. The car, a bread-and-butter sedan festooned with "no emissions" logos, was not much to look at, despite its status as one of the world's very few operational fuel-cell cars.

Just by opening the doors, though, the P2000's lightweight status becomes clear. I'd been waiting to drive a fuel-cell car for a long time, so it was a revelation to point this one down the track and hit the accelerator. I'd heard of a fuel-cell "lag" like that of early turbocharged cars, but the Ballard-equipped Ford, with the equivalent of 90 horsepower, provided a satisfying surge of power. In that stage of development, the car could reach sixty miles per hour in fourteen seconds, with a top speed of ninety. It was delivering one hundred miles on a tank of hydrogen, but the engineers said it would soon go much farther. It was noisy and unfinished, but that's hardly unusual for a prototype.

Will people buy a high-efficiency P2000? The recent marketplace performance of cars such as the Chevrolet Metro and the Dodge Neon have not been encouraging, and a highly fuel-efficient version of the Honda Civic, the fifty-nine-miles-per-gallon VX, was a flop in the United States. Blame it on cheap gas. Like many of his Detroit colleagues, Wallace thinks the Prius is a brilliant car—for the Japanese market. "When you drive it here, because of all the highway driving, it gets only forty-two miles per gallon. Ford will build a hybrid car, absolutely, but it will be a hybrid suitable for our major customers." Ford, and GM too, may build hybrid sport-utilities. If Ford could build an Expedition or Excursion that did everything the current car does but gets great gas mileage, it's hard to see how it could go wrong.

The process of cleaning up the sport-utility has already begun. In early 1998, Ford stunned its competitors by announcing that its Explorers and Expeditions would henceforth meet the California low-emission-vehicle standard. That decision had the imprint of company chairman William Clay Ford Jr., the committed environmentalist who is the first family member to serve in the company leadership since the days of Henry Ford II. The younger Ford has alarmed some financial analysts who fear, as *The New York Times* put it, "that the scion of a billionaire family could put environmental causes ahead of profits and undermine the industry's traditionally united front against pressures from environmental groups."[15] In late 1998, *Newsweek* put Bill Ford on its cover, riding with him in a Ranger pickup EV and listening as he predicted a day when "a large portion of society is going to be driving alternative-fuel vehicles."[16] Ford backed that prediction with cash a few months later by acquiring a controlling interest in Norwegian EV producer PIVCO. Ford will sell the small plastic-bodied PIVCO electric city cars in the United States as part of its effort to meet the 2003 California mandates.

Bill Ford has to reconcile two widely divergent missions, cleaning up the company and keeping it profitable. Sometimes these warring impulses surface simultaneously, as in a 1998 Dearborn speech in which he proclaimed both that his interests were "fully aligned with those of all shareholders" and that he wanted Ford to become "the world's most environmentally friendly automaker." It may not be easy to have it both ways. Ford's best-selling but gas-guzzling Expeditions and Lincoln Navigators are also its profit center, earning the company as much as $15,000 each. From just one Wayne, Michigan, factory making sport-utility vehicles, Ford earns approximately $3.7 billion a year, enough money for the company to pay for its purchase of Volvo in less than three years.[17]

Environmental groups have been appreciative of Ford's public statements, but they want action, too. In February 1999 a coalition of twelve groups, including Friends of the Earth, Greenpeace, the Sierra Club, and the American Lung Association, sent Bill Ford a letter saying that they were "encouraged" by what he'd said and done so far, but they wanted him to go further by endorsing stringent Tier 2 emissions standards, a radical departure from the auto industry's business as usual.

They also asked for a face-to-face meeting. "If Ford really is serious about casting itself as a green company, it has to step up to the plate and announce, 'By God, we're going to do it,' " said Frank O'Donnell, executive director of the Clean Air Trust.

The meeting didn't happen, partly because of bad timing. Even as Ford was issuing green press releases, it was also making clear its ongoing duality by introducing, within weeks of the environmentalists' letter, the huge four-ton Excursion, a "high-end" sport-utility vehicle priced at $45,000 to $50,000. The nineteen-foot truck gets only twelve miles per gallon, and environmentalist groups reacted angrily to this affront to clean air. Dan Becker, the Sierra Club's director of global warming and energy programs, called the Excursion "a rolling ad for improving auto-pollution standards."

Ford officials, who seemed to expect that Bill Ford's appointment would at least entitle them to a temporary cease-fire, were disgruntled. They pointed out that the Excursion, a rival to the Chevrolet Suburban, is cleaner than it has to be and qualifies as a "low-emission" truck in California. Instead of agreeing to a meeting, or committing the company to endorsing stronger air-quality standards, Ford replied personally to all the signatories of the February 5 letter, reiterating his intention "to make environmental stewardship part of the mainstream of our business."

For his part, Chris Ball, outreach director at the global-warming crusader Ozone Action, one of the signatories to the letter, wants to see bolder steps from the company and less green-sounding rhetoric. Ball, who met with Bill Ford as a member of a delegation from the Interfaith Center on Corporate Responsibility, said, "My sense is that he's a good man in a tough position, trying to move an awfully big corporation. The environmental community is watching him very carefully." Ball praised Ford's environmental track record, but he insisted that "the good work is dwarfed by the company's continued insistence on building behemoths like the Excursion. If you look at the *Green Guide* list of the twelve worst vehicles for the environment in 1999, three of them are Fords."

Ford is a conglomerate of wildly divergent parts, and the gas-guzzling Excursion sits alongside zero-emission EVs in company parking lots. After my meeting with Wallace, publicist Brendan Prebo handed me the keys to a bright blue Ranger pickup, an EV conversion that is just like the one Bill Ford drives. Only 450 of these trucks have reached customers since the vehicle was introduced in 1997; most are in the hands of utility-company fleets. Ford has made little effort to market the Ranger to the general public, and, given its low expectations for battery-powered EVs, the situation isn't likely to change.

I drove the Ranger around Dearborn and found it quite advanced, fast, rattle-free, and quiet as a church mouse. But could I see a battery-powered truck like this in mass production, with a good old boy from Texas behind the wheel, happily adjusted to the one-hundred-mile range? Not quite.

CHRYSLER'S INTERNATIONAL OUTLOOK

When it comes to EVs, Chrysler has historically been America's reluctant advocate, though its merger with fuel-cell pioneer Daimler-Benz has jump-started interest at the number three automaker. The company's environmental advocates have to overcome a history of skepticism.

New model introductions are usually upbeat affairs, but François J. Castaing, then Chrysler's vice president of vehicle engineering, looked positively glum when he announced plans for the Electric Powered Interurban Commuter (EPIC) minivans at the company's Auburn Hills, Michigan, headquarters in 1994. "Building them is not the issue," Castaing said. "Selling them is the issue." He likened Chrysler's own EPIC transporter to a gasoline van "with a two-gallon fuel tank, an orifice that big"—he formed his fingers into a small circle—"and no place in California to fill it up."[18] Castaing informed the press that the van's performance and range would not begin to satisfy drivers used to internal combustion.

Chrysler's pessimism was also reflected in the company's public testimony before bodies like the California Air Resources Board. In 1995, when the board was holding hearings on its 1998 mandate, William Glaub, the company's general sales manager, told the regulators, "I am particularly concerned about the very high initial cost, unacceptably short range—even on a good day—and the need to plan for all-night recharging. . . . *With all of the other choices out in the marketplace, electric vehicles will be difficult to move* [Glaub's emphasis]. *My recommendation is to scrap the mandate.*"[19]

And, of course, the mandate was scrapped—at least for 1998. But the 10 percent mandate for 2003 remained in effect, so Chrysler reluctantly geared up for EVs. Chrysler has been a very adaptable company, with a reputation for engineering innovation and styling leadership that was somewhat marred by quality-control problems. CEO Robert Eaton and president Tom Skallkamp revived the company from the doldrums it was in at the end of the Lee Iacocca era, but EVs had always been an afterthought. The company fielded a tiny fleet of fifty electric TEVans to selected utility customers in the early 1990s, and its battery EV programs have not grown significantly since.

Introduced for the 1997 model year (when a grand total of eighteen were leased), the EPIC van was competent but no great shakes. Featuring a 100-horsepower AC-induction motor and a rack of lead-acid batteries, the van could reach eighty miles per hour but had a range of only sixty miles. (When used in cold upstate New York, said Chrysler's Mike Clement, the range dropped to only fifteen miles.) Echoing its belief that the general public just wasn't interested, Chrysler made the vans available only to government and utility fleets. Obviously, it wasn't going to change the world.

In 1999, the company, now called DaimlerChrysler, put nickel metal hydride batteries into its minivans, and the range climbed to ninety miles. For its battery packs, DaimlerChrysler made an end run around Stan Ovshinsky and GM-Ovonic by buying from French battery maker SAFT, which set up a manufacturing site in Georgia. The vans got much better with the advanced batteries, and two thousand were built in 1999, but they were still available only to fleets in New York and California.

DaimlerChrysler has more interesting projects than its EPIC vans. In 1994, it built a fascinating natural-gas-powered hybrid race car called the Patriot, to compete in World Sports Car endurance events like the French twenty-four-hour Le Mans race. The car was actually put together in England by Reynard, a well-known Formula 3 fabricator. Under the hood of the Patriot was a twin-turbine turbo-alternator, which was coupled to a rapidly spinning, power-storing carbon-fiber flywheel to form what's known as an electromechanical battery—similar in concept to the flywheel system developed by Rosen Motors. Westinghouse built traction motors for the car, theoretically giving it a top speed of two hundred miles an hour. It recalled the heady days of Camille Jentzy's La Jamais Contente, a streamlined electric racer that could hit sixty-five miles per hour in 1895.

The Patriot was fiercely complex, technically advanced, and, unfortunately, a nonstarter in racing events. Chrysler didn't immediately follow up the hybrid design, and a company brochure of the period explains why. "Have hybrid vehicles ever been successful in the automotive marketplace?" it asked. "No," it answered. "Hybrid technology is expensive: there are two 'power' units, the fuel-burning engine and the electricity-driven motor." Internal combustion, it said, was simpler and more flexible.[20]

Still, DaimlerChrysler's ESX2, which it debuted at the Detroit Auto Show in early 1998, was a very well thought out, lightweight, and aerodynamic seventy-miles-per-gallon hybrid car, and (like Ford's P2000 and GM's EV1) an excellent platform for a range of alternative-fueled vehicles, including a fuel-cell version. What's more, Chrysler's merger with Daimler-Benz put it in a position to benefit from all the advanced work coming out of the Alliance. The potential was there, and that's what led me to Chris Borroni-Bird.

Borroni-Bird, who has a Ph.D. from Cambridge, seems more like a young British academic than DaimlerChrysler's manager of technology strategy. His office is tucked away in a small DaimlerChrysler warren in Madison Heights, miles from the central headquarters in Auburn Hills. When I arrived, a group of shirt-sleeved employees stood in the parking lot admiring a gaudy, pollution-spewing dirt bike parked atop an extended-cab Dodge V-8 truck. The modest world of fuel cells and hybrids seemed far away.

In fuel-cell circles, Borroni-Bird has been a persistent advocate of gasoline reforming. He has argued that staying with a proven fuel will make fuel cells happen much more quickly, and that marketplace forces would then push the switch to cleaner sources. "I think they hired him to make that case," said Jim Cannon, author of the book *Harnessing Hydrogen* and president of Colorado-based Energy Futures, Inc. "I've seen him at twenty conferences, and he was always advocating using conventional fuels like gasoline and diesel in fuel cells." François Castaing was equally blunt. At the Detroit Auto Show in 1997, he said, "We believe hydrogen needs to be processed from gasoline on-board vehicles because hydrogen isn't a practical fuel choice today."[21]

Borroni-Bird denied that he ever had a hidden agenda. "The approach DaimlerChrysler takes is fuel-flexible," he told me. "People think we're anti-methanol, which is not the case. Politics often decides what fuel is used: ethanol, for instance, can be made from corn and is popular in the Midwest. We're talking about an onboard sensor that can tell what kind of fuel is being poured into it, then adjust the reformer on the fly."

Such a system would have to be quite complex, since different fuels are reformed at different temperatures, using varying proportions of steam and air. Almost no other company endorses this approach, though most agree that reformers should be designed to be easily adaptable for the wide variety of fuels in use internationally. (Brazil would get ethanol reformers, for instance, and America methanol or gasoline models. Most countries have a single dominant fuel.)

Borroni-Bird ticked off all the disincentives for carrying onboard hydrogen: It leaks easily, is hard to store and hard to compress, and burns quickly. Even if it could be safely carried, it may be hard to get costs down. Refueling is apt to be difficult. But even given all that, Borroni-Bird was quick to assert, "The fuel cell has such profound advantages, it's worth pursuing aggressively. And, with Daimler-Benz, we are pursuing it aggressively."

There are three central automotive goals, he said, efficiency, range, and emissions. "Diesel has the efficiency and range, but there are emissions problems. Batteries have the emissions and the efficiency, but not the range. The fuel cell promises to have extremely low emissions, *with* excellent range and efficiency. Hydrogen is an amazing substance. It's lighter than air. In its liquid form, you could throw it at people and it would evaporate before it hit them. Fuel cells are like a dream, aren't they?"

A rather slow dream. Dr. Borroni-Bird offered one of the more pessimistic timetables coming out of Detroit. He thinks fuel-cell stacks are feasible, but not in the short term. "I'd say that 2010 is a date that I could be comfortable with, another ten years of R and D," he told me. Complex fuel processors that can handle gasoline, like the system developed by Arthur D. Little, "have been proven to work," he said, "but actual production models are still a long way off." In fact, DaimlerChrysler's efforts to make a gasoline reformer work have been so disappointing that, at the 1999 Los Angeles Auto Show, the carmaker announced it would henceforth concentrate its efforts on methanol, signing on to the program advanced by its partners in the Alliance. Ironically, DaimlerChrysler showed off a gasoline-powered fuel-cell Jeep at the same show, but the car represented a developmental dead end. Soon after, the Alliance partners announced an ambitious testing program in, of all places, Iceland. It will start with a small fleet of fuel-cell buses in the capital, Reykjavík, then slowly convert every vehicle on the island—even the ubiquitous fishing boats—thus creating the world's first hydrogen economy.

Borroni-Bird is somewhat jaundiced about hybrids, too, since "you can't get Americans excited about fuel savings. The cars we've seen, like the Toyota Prius, are heavily subsidized by their makers, and I'm not sure they become affordable at higher volumes." No one could accuse the man of making rosy predictions.

At least until the merger with Daimler-Benz, Chrysler was working diligently on both fuel-cell and hybrid technology, though the Germans are now expected to take the lead. The innovations built into the ESX2 will undoubtedly surface in several production versions. Getting a gasoline reformer to work efficiently in a moving car will be a much more daunting proposition, though that would move the timetable for the practical fuel-cell car forward. In the meantime, Daimler-Chrysler will sell you an EPIC minivan. If you're a fleet operator, that is.

The Detroit show cars I've had a chance to inspect are like Potemkin villages—they look good from a distance, but inside they're half-finished and usually nonfunctional. After the crowds have gone home, they're often dismantled or crushed. Are these new EV prototypes on a similar ride to nowhere? I don't think so.

Instead of working alone on EVs, as has been their pattern, the Big Three have joined in global alliances—with Daimler-Benz, Volvo, Mazda, Toyota, and others—that have already yielded impressive results. Even if Detroit's heart is still with internal combustion, practical and competitive forces are forcing it to eventually abandon that familiar technology.

And a new generation of leadership is emerging. The importance of William Clay Ford's ascendancy to the chairmanship of the company that bears his name can't be overstated. This is an auto executive who uses organic lawn fertilizer and convinced the company to print its business cards on recycled paper. More substantively, he signed up environmental architect William McDonough to redesign the ancient Rouge plant in Dearborn (built in 1916) and turn it into a model of green efficiency. Bill Ford thinks that environmental consciousness has to become fully integrated into the corporate structure. "If all this is just [me] waving the green flag, saying, 'We need to get cleaner, guys,' it's not going to work," Ford says. "It's going to work if we embrace it in the mainstream and make it part of the way we do business."[22] Connected through the Alliance to like-minded DaimlerChrysler and Volvo, Ford is emerging in a strong position to lead the industry forward. The big question is, can such leadership triumph over bottom-line concerns?

It would be facile to think that Detroit would build clean cars because "it's the right thing to do." Ultimately, underengineered vehicles half-heartedly designed to meet a mandate would fail in the marketplace. But the gasoline engine can be compared to the VHS videocassette, an obsolete analog technology whose global distribution and supporting infrastructure kept it on the market while the industry squabbled about a digital successor. Carmakers are beginning to think they can build not just a cleaner car, but a better one.

I went to Detroit in 1998 hoping that the industry's energetic lobbying against green initiatives didn't tell the whole story, and I came back convinced that transformation is in the air. Maybe it's going too far to call it a revolution, but "evolution" is certainly fair. I noted with interest that the American Automobile Manufacturers Association, the strident anti-environmental voice that led the fight against the 1998 California mandates, was disbanded at the end of 1998. In its place is a new and less polemical international body called the Alliance of Automobile Manufacturers (AAM), representing nine companies (including all of the Big Three). Peter Pestillo, AAM's first leader and a Ford vice chairman, announced a "special commitment" to environment issues. Is that just industry window dressing? It could be. One of AAM's first battles, in the fall of 1999, was a campaign to freeze CAFE standards in place for another year.

CHAPTER 6

THE GLOBAL GREEN CAR: GERMANY AND JAPAN ON THE FAST TRACK

IN THE LATE 1960S, I held a high school job in a Dodge dealership that had taken on the unenviable job of trying to convince Americans to buy Japanese Toyotas. There was little sense that the tiny Corollas I picked up and prepared for sale, derided as toys by the older salesmen, were about to set the American auto industry on its ear.

After a decade of U.S. auto industry resurgence, the imports may be at it again, delivering fuel-efficient, environmentally friendly automobiles to a public that won't be able to get what it wants from Detroit. The prospect of that happening again as it did in the 1970s is what keeps the U.S. automakers interested in alternative fuel. I realized that the clean-car story was not solely an American one, and that I'd have to get my passport renewed to tell it properly.

In Japan, all the major manufacturers are developing hybrid and fuel-cell automobiles, aided by a supportive central government that pours money into solar power and basic hydrogen research. In Germany, DaimlerChrysler has mounted a strong push on fuel cells and will probably be the first company to bring them to market. The French are leading the way to the eco-city of the future with small urban EVs and systems that turn them into a sharable form of public transportation.

In surveying the international scene, I was interested to see how the EV would fit into everyday life, knowing the hurdles it faces in the United States, where energy waste and big-car indulgence is practically a birthright. The unexpectedly strong sales of the Prius on its 1997 debut in Japan provides a partial answer, as do the fuel-cell taxis plying the streets of London. The eco-consciousness that thrust the Green Party into a leadership role in Germany is widespread in Europe and evolving in Asia. Make no mistake: Internal combustion still dominates the world, though its grip is destined to loosen.

DAIMLERCHRYSLER: FUEL-CELL LEADERSHIP

If arriving passengers make it through customs and passport control at the airport in Stuttgart, Germany, they can walk right through the portals of a high-speed light-rail system that will whisk them into the center of the city in about half an hour. At the central train station, which shares a complex with the light-rail system, expresses are waiting to take them all over Europe. Sleeping accommodations are available.

With public transportation like that, it's sometimes hard to understand why Germans are so in love with their cars. In Stuttgart, where DaimlerChrysler has its world headquarters, private cars zip through narrow city streets at breakneck speeds. There's no obvious smog, but acid-rain damage is easy to spot on the pine trees fringing the highways.

Its partners in the Alliance are Ballard, Volvo, and Ford, but DaimlerChrysler seems the senior partner and has been undertaking most of the more visible vehicle testing. The company has been only too happy to take the public for rides in the succession of car and bus fuel-cell prototypes it built, beginning with a hydrogen-powered internal-combustion minibus in 1975. In the 1980s, before its significant work on fuel cells, the company (then Daimler-Benz) was intrigued with the idea of burning hydrogen in internal-combustion engines, and it conducted a large-scale road test in Berlin from 1984 to 1988, involving ten vehicles and more than 350,000 miles of driving.[1] (In the same period, BMW also began testing cars that burn liquid hydrogen, an experiment that continues to this day.)

The German government committed more than $100 million to these projects, which inevitably languished because burning hydrogen is inherently less efficient than splitting it chemically in a fuel cell. And combusting hydrogen indirectly releases emissions of carbon monoxide, hydrocarbons, and particulates, albeit only about a tenth of that resulting from the burning of fossil fuels. Daimler-Benz was pursuing the twin holy grails of high efficiency and zero emissions, and its research finally bore fruit when the company built NECAR (New Car) I, a commercial van that was its first fuel-cell vehicle, in 1994.

The technology moved forward rapidly from there. NECAR I was a van in name only: Aside from the driver and passenger, the entire vehicle was full of fuel-cell apparatus, and the roof held an enormous hydrogen tank. By the time of NECAR II, a smaller van, in 1996, seating for six was restored. NECAR II can hit sixty miles per hour and cruise for 150 miles before its onboard hydrogen tanks are refilled. But was 150 miles enough? The range problem got Daimler-Benz into the quagmire known as fuel reforming, as seen on NECAR III. With an onboard reformer, the fuel-cell car's potential range zoomed up to three hundred miles or more.

In the past, Daimler-Benz might have been content to develop the technology slowly, releasing a few prototypes to soak up light from the press's flash units. But by the late 1990s, a wholly more impatient breed had taken the helm of the auto business. At Mercedes, Daimler-Benz's car division, a penchant for glacial change and overengineering came to an abrupt end in 1993, when the refreshingly candid Jurgen Schrempp (known in Europe as "Neutron Jurgen" or "Rambo") took over. Mercedes in the 1980s had been, if not smug, at least supremely confident that its technical excellence would sell itself.

At one Mercedes press event I attended in those years, company spokesmen calmly announced themselves unworried by the then-impending introduction of luxury rivals from Japan such as the Lexus LS400 and the Infiniti Q45. But soon it became apparent that Tokyo could design and build cars that were 90 percent as good at 70 percent the price. The market for $150,000 V-12 S600 luxury sedans with double-glazed windows began to shrink. "For years we had no competition in engineering quality," one executive says. "But now our competitors are getting better and better."

In the mid-1990s, with Schrempp at the wheel, Daimler-Benz pushed itself out of a narrow niche of its own devising and began entering markets it formerly ceded to others. It finally released a sport-utility vehicle, designed a new affordable sports car that sold for $40,000, a price it would previously have scoffed at, and even entered the European small-car sweepstakes with its tiny A-Class—and took 100,000 advance orders. Even more dramatically, it put the Mercedes star on the $9,000 two-seat urban Smart car, a golf cart with license plates. Clearly, Mercedes was thinking about the future, and about fuel cells.

NEBUS, which arrived in 1997, reflects the downsizing work done by Ballard and carries ten of the company's 25-kilowatt fuel-cell stacks in a neat rear compartment. It's an altogether normal-looking and completely functional city bus, with comparable range, similar but not identical to the buses Ballard has cruising the streets of Vancouver and Chicago.

In 1998, Daimler-Benz delivered a significant breakthrough, the world's first methanol-reformed fuel-cell car. As if challenging itself, Daimler used its subcompact A-Class as a base. NECAR III was not a polished product: It was a heavy, rolling test bed, and the fuel-cell stacks and reformer took up everything aft of the front seats. Unlike the hydrogen-fueled cars, it didn't perform smoothly. But it ran, and it certainly proved the concept. A methanol fuel-cell car was viable.

In May 1998, an energized Daimler-Benz announced that it was merging with Chrysler, creating the fifth-largest carmaker in the world, though the terms of the deal made it look more like an acquisition. Daimler-Benz's shareholders received 57 percent of the stock in DaimlerChrysler, and Chrysler's 43 percent. Chrysler head Robert Eaton co-chaired the management board, but his retirement was only two years away.[2]

Shortly before the DaimlerChrysler announcement, I visited Stuttgart for an international press gathering at the Mercedes Forum downtown and listened as company executives talked about technological innovations. I was intrigued by the new generation of cruise controls, introduced on 2000 models, which automatically sense when your vehicle is getting too close to the car in front of it and smoothly back you away. With modern global positioning, it seemed, you would never be lost in Tokyo or New York.

We were also given brief rides around the convention center in the fuel-cell vehicles NECARs I and II, and the NEBUS. Working fuel-cell vehicles are hard to find, and these were the first I'd seen up close. It was a little bit anticlimactic. Riding in these conventional-looking vans and buses, with only an air compressor's whine to give away their identity, I didn't feel like George Jetson. Couldn't they have hovered an inch off the ground?

Conspicuously absent from the press demonstration was NECAR III, which was obviously not yet ready for international display. I wanted more than a brief overview about the company's fuel-cell program and a ride around the block, and it became clear I'd have to come back for a more focused visit. And so, in August 1998, I ventured out to Nabern, home of the Fuel Cell Project House, where highly technical work is performed in the shadow of apple orchards and a looming castle.

The silver-maned Ferdinand Panik runs the Fuel Cell Project House, which has about thirty employees of its own and shares space with Ballard's German arm and a joint venture called Daimler-Benz-Ballard (DBB). Where other executives are full of cautions, Panik seems supremely confident that he can meet the industry's most pressing timelines. Not only will DaimlerChrysler have production-ready cars by 2004, he said, but it will also have production itself, as many as forty thousand fuel-cell cars that first year. "It's feasible," he told me. "We have a schedule, and we are sticking to it." Panik envisions a two-track infrastructure, with pure hydrogen gas for fleet vehicles, everything from delivery trucks and taxis to buses, and onboard reformed methanol for passenger cars. The plan is nothing if not practical, since fleet vehicles are usually served by large garages with trained staff and the facilities for in-house hydrogen production. Without reformers, fleet vehicles could be technically less complex and would be able to work within the 250-mile range limitation posed by the carrying of hydrogen tanks.

Talking with Ferdinand Panik, I realized why DaimlerChrysler hadn't handed over the keys to NECAR III during my previous visit: He wasn't satisfied with it. "It's our first methanol vehicle," he said, "and the reforming technology is very complex. It takes time for the four-step process to work, so there is hesitation when you accelerate, plus the problem of excessive noise from the compressor."

Unfortunately, I didn't get to hear that excessive noise since, as I'd come to expect with experimental cars, NECAR III was in pieces when I finally saw it. The engineers were trying to address that hesitation problem with a new injection system for mixing methanol and water. The car, festooned with environmental decals, sat on a hoist, without headlights or bumpers, its rear-compartment covers lifted off to reveal its complex inner workings. The fuel cells themselves, two Ballard 25-kilowatt stacks, sat on a rack. Although it's a small car, it was virtually handmade (by Ph.D.s!) and represents a big investment. Gerald Hornburg, the DBB fuel-cell system manager who is one of the car's keepers, estimates NECAR III to be worth more than $5 million. Pretty good for a car with a hesitation problem!

Panik is convinced that hydrogen is the fuel of the future. "No technology lasts forever, and it is time to replace fossil fuels," he said. "I believe hydrogen offers the best opportunity to do that, and I don't see anything else coming along with the same potential. Fuel-cell research is becoming a major international trend. It's amazing how many engineers are working on this technology, not only here but worldwide. I think it's become a matter of when it will happen, not whether it will happen."

Panik's conviction that methanol will go into fuel-cell tanks arises from the fact that, according to DaimlerChrysler, Germany already has enough methanol production to fuel 100,000 cars. Worldwide, there's sufficient methanol for two million cars, an impressive figure but nonetheless smaller than the existing gasoline market in Germany.

DaimlerChrysler has renewed its interest in liquid hydrogen, and that was the fuel in the version of NECAR IV that the company unveiled in a March 1999 Washington ceremony, with Environmental Protection Agency Secretary Carol Browner on the podium. DaimlerChrysler calls NECAR IV "the first drivable, zero-emission, fuel-cell car in the United States," and while that claim could be disputed (because Ford and GM also have running fuel-cell prototypes), there's no question that it represents a huge advance over NECAR III, whose cell and reformer took up all the passenger space. NECAR IV was still heavy and slower to accelerate than Ford's P2000, but it could boast room for five, with a 40 percent power increase over the earlier version, a higher top speed (ninety miles per hour) and a range of 280 miles. The numbers looked good, but liquid hydrogen is a very difficult fuel, and no one is predicting that the first DaimlerChrysler fuel-cell production cars will actually run on it.

The company is intrigued by the liquid fuel's easy portability, as is BMW, and both are in partnerships with the German company Linde, which builds liquid hydrogen refueling stations. But handling liquid hydrogen is a high art, since hydrogen reaches a liquid state only at minus 400 degrees Fahrenheit, and a single drop of the supercold fuel will cause serious damage to skin. Liquid hydrogen "gas stations" would probably have to be run by robots, and there already is such a station in Munich. Making liquid hydrogen work requires some seemingly bizarre solutions, like attaching the tank to the car with magnetic holders to isolate it from thermal convection. But a liquid hydrogen tank need not be much bigger than a gasoline tank, and it would offer comparable range.

According to Johannes W. Ebner, a Project House vice president who works closely with Ferdinand Panik, a superinsulated tank could keep liquid hydrogen cold for weeks, but if left untended for longer than that it would inevitably warm up and return to a gaseous state, requiring that it be vented from the tank (a harmless process). Might that hydrogen gas be recaptured and reused by the car? Such are the challenges keeping the Project House humming.

Ebner puts the hydrogen fuel cell in strategic context. "Some seventy percent of the world's oil is in OPEC countries, and sixty-five percent of it is in the Persian Gulf," he says. "That's a big dependency. If we can switch to methanol, which is produced from natural gas, we can get over that dependency. Here in Germany, our natural gas comes from Russia and the Netherlands, and from many other places, too. We think oil production will eventually reach a peak, and then start to decline. Within the U.S., it's already past the peak. In the North Sea, it will peak in the first half of the next decade. In the Gulf, that point will come in the next ten to fifteen years.

"We are facing more demand in the Third World. But China is on a path to avoid a big dependence on crude oil. They have coal and natural gas, which they could convert to methanol, or use natural gas directly. Hydrogen fuel cells make a lot of sense for China, but it's an expensive technology requiring considerable investment. A Chinese delegation recently came here and spent four or five weeks learning about our vehicles, with a particular interest in fuel-cell buses. We're in the first stages of learning from each other."

The progress DaimlerChrysler made in just a few years woke up the world to the possibility of fuel cells. If NECAR III was a revelation, NECAR IV, with the fuel-cell and methanol processor taking up a quarter of the space, was even more of one. In this new technology with a 160-year history, DaimlerChrysler is determined to be the first in line. "We Europeans missed the competition on watches," Ebner says. "They were developed in Switzerland, but now they're all built in Japan, China, and the Philippines. In fuel cells we want to be on top. It's a competition-driven international race, and the company that is first to production will get to make the rules."

Before leaving Germany, I traveled out to the nearby industrial center of Sindelfingen, where Mercedes-Benz makes C- and E-Class cars and excited Americans come by for European delivery. It was easy to get overwhelmed by the numbers. According to guide Uwe Ankele, the factory has eighty buildings, thirty thousand employees, and produces eighteen hundred cars a day, using up more than sixty miles of sheet metal (delivered just in time for the stamping machines) in the process. It is a highly robotized assembly line, with far fewer workers than you'd see in an American plant. I watched a precisely articulated mechanical arm engraving chassis numbers on a body, and another fitting in windshields in ten precise movements, after gently applying glue around the perimeter. Though Sindelfingen had the huge scale and deafening industrial din of every other auto plant I've been in, there was some sense of the company's environmental concern. Mercedes workers use bright yellow in-house bicycles to get around the huge factory, and the company makes its own electricity in a clean-burning 1.4 million–kilowatt natural-gas plant.

Outside the factory were rows of cars waiting to be shipped out. Sixty-five percent of the production is exported, and in three weeks these cars would be in Bangkok or New York. Could this industrial behemoth change direction and begin dropping fuel cells into engine bays that once held internal-combustion engines? Given the changes Daimler-Benz has made since the late 1980s, it seems entirely possible.

AN IMPATIENT JAPAN'S ACCELERATED SCHEDULE

Most new car announcements are made at one or the other of the international auto shows, a peculiar phenomenon that marries technical innovation to Hollywood glitz and blatant sex appeal. Although some "liberated" carmakers tried briefly to introduce male "spokesmodels," they were quickly found wanting and banished. Today, nearly every car stand features a young woman chanting a memorized mantra. TVR even put a pair of nude women on its London stand in the 1970s. The effect on sales was not recorded.

The 32nd Tokyo Motor Show in October 1997 turned the

auto industry on its head and pushed a company better known for aggressive marketing and near-flawless quality control into the forefront of the green revolution. The company was Toyota, whose chairman and former president, the demanding Hiroshi Okuda, had made his priorities plain. "Consumers are smart," he said. "They recognize the threat that pollution and global warming present to them and their children. Our job is to present [them] with a broad selection of clean, green technologies." He said that Toyota would give "top priority" to reducing auto emissions.[3]

Company chairmen say that kind of thing a lot without doing much about it, but Okuda backed up his words, not with mere prototypes but with real production cars. The prime example was the Prius, a hybrid car with both an electric motor and a small gasoline engine that had its genesis in a 1995 show vehicle, and it got everyone's attention in Tokyo. The carmaker actually put this little Corolla-like sedan, the world's first modern hybrid car, on the Japanese market in 1998, for an eye-popping $17,000. And it immediately began to sell well.

If it had been, say, Subaru, introducing the Prius, the introduction might not have been so momentous. But Toyota was different, a top contender with an uncanny knack for sensing changes in wind direction. "I don't know of a company that better combines superior skills in all the critical areas: manufacturing, engineering, and perhaps marketing," said Michael Cusumano, a professor at MIT's Sloan School and a participant in its International Motor Vehicle Program. "If they wanted to blow away GM, they could."[4] A significant percentage of early Prius production models went to other car companies for intense evaluation. GM bought four, Chrysler three, BMW and Honda one each.

Unlike its American counterparts, Toyota wasn't fighting the bad news about global warming. In fact, it was joining in an industrial alliance, with British Petroleum and others, to fight carbon dioxide buildup in the atmosphere. It was donating money to research centers looking into the problem and was putting its own operations through a top-to-bottom environmental restructuring. "For a company to benefit from 'green' concerns, a serious shift to an environmental commitment must be made at the top," says a company periodical. "This is exactly what has happened at Toyota."[5]

In May 1998, Toyota's U.S. sales operation became the first California-based company to buy 100 percent renewable energy certified under the Center for Resource Solutions' Green-*e* program.[6] The company also won the CALSTART 1997 Blue Sky Award, recognizing its work not only on the hybrid Prius but also on the Coaster hybrid electric bus, the RAV4 electric car, and its unrelentingly cute e-com two-seater electric commuter car.[7]

While other companies are planning to introduce hybrid cars, Toyota has actually done it. By the summer of 1999, more than twenty-five thousand Priuses had been delivered to Japanese customers. Monthly production jumped initially from one thousand to two thousand, triple the original sales projections. The automaker brought a modified Prius into the U.S. in 2000, with ambitious sales targets of thirteen thousand a year, more than half of the total global production for the international market. The Prius's low sales price is heavily subsidized on the Japanese market, and it may well be in the United States, too, if economies of scale can't reduce the automaker's costs. Toyota is slotting the U.S. car between the $15,500 of an average Corolla and the $22,500 of the Camry.

To prepare the Prius for U.S. sale, it had to be adapted to U.S. driving conditions, which include a lot of freeway mileage. U.S. emissions standards had to be adhered to, and there were changes to engine torque curves and hot-weather performance. Prius engineers tested the car in Death Valley, because Japan has no similar high-temperature, low-humidity conditions.

The Prius team also took into account American fuel prices, a third of those in Japan. American critics point out that the cars' much-hyped sixty-six-miles-per-gallon fuel economy will be sharply reduced if owners use them mostly to commute on limited-access highways, where the added weight of the electric-drive system just slows them down.

To ease the Prius's way in the United States, Toyota took six of the Japanese-spec cars on a thirteen-city American tour in the summer of 1998. Although the tour covered many major cities, from Los Angeles and San Francisco to Atlanta, Miami, and Boston, it skipped New York City in favor of Albany. Why Albany? Because that's where New York's state legislators are, and Toyota was responding to Governor Pataki's contentious EV mandate, which required the major auto players to start selling zero-emission cars in the state. By the time Toyota learned that a federal judge had struck the mandate down (in the same month as the tour), it was too late to change the venue.

Toyota's fuel-cell efforts are the source of endless speculation in the West, and the company is characteristically tight-lipped about its plans. Toyota's work with PEM cells began with a feasibility study in 1989. It showed a drivable methanol-reformed "FCEV" fuel-cell car, based on its electric RAV4, at auto shows in Frankfurt and Tokyo in 1997 and has demonstrated interest in storing hydrogen in metal hydride, a technology most other companies have tried and rejected because the metals are too heavy. But Toyota says it can obtain a 155-mile range with metal-hydride storage. Toyota will share some of its fuel-cell research information—and technical data from some of its other environmental initiatives, including vehicle recycling and reduction of greenhouse gases—with GM, its partner on a number of projects.

Toyota has given some mixed signals. In July 1998, Toyota vice president Akihiro Wada said the company would try to have a fuel-cell automobile ready by 2003, but three weeks later then President Okuda downplayed that prediction, indicating that "the prospects for his company to develop and market commercially viable fuel-cell vehicles by 2003 are not as positive as initially thought."[8] In the spring of 1999, Toyota announced it would share its technology with occasional partner GM in a five-year collaboration on electric, hybrid, and fuel-cell cars, probably speeding up the EV timetable for both companies.

Indications of where Toyota is going are best gathered from auto industry veterans like Dave Hermance, a 1960s graduate of the legendary General Motors Institute. But while most such graduates stay at GM for their entire careers, Hermance lasted only twenty-six years, until 1991, when he joined Toyota as general manager of Powertrain at the Technical Center in Gardena, California. "The Japanese auto culture is very different," says Hermance, who proved his "car guy" credentials by dropping a small-block Chevy V-8 engine into a Triumph TR3 and driving it all out. "The Japanese have a propensity to fix every problem they find," he says. "At GM, they're more likely to say there's something wrong with the test."

On fuel cells, Hermance says that Toyota "seems to be retrenching its position" about a delivery date. He says that, as of 1998, the research division was testing both methanol reformers and metal-hydride hydrogen storage, and had two prototypes of each design. "I'm not sure what system we'll end up with," says Hermance, "but my personal belief is that we'll never solve all the cost problems for onboard reformers. Direct hydrogen, however, is doable by 2004—with some luck. It's a huge technical challenge."

Hermance believes the company's commitment to environmental change is real, not window dressing. "Toyota was convinced even before the Kyoto global warming summit that we have to minimize our footprint on the overall ecosystem. And the company is looking at all kinds of things to improve fuel economy and hold down emissions."

Toyota is not the only Japanese automaker thinking green thoughts. Mazda, which is partially owned by Ford, is working with the Alliance on fuel-cell cars, with a target date of 2003 or 2004, and also has its own interesting experiments of long standing, including a rakish hydrogen-burning rotary-engined Miata. Nissan plans to introduce an aluminum-bodied hybrid car soon, though not in the U.S. market. Also a Ballard customer, financially troubled Nissan intends to have a methanol fuel-cell car on the market between 2003 and 2005 and began leasing a lithium-ion battery-powered station wagon called the Altra EV to public agencies in California in 1998. Mitsubishi is also marketing a hybrid in the United States.

Honda, whose hybrid seventy-miles-per-gallon two-seater Insight (based on the J-VX and VV show cars) could well be a hit in the United States, has taken an unusually open approach to clean-car technology. In an interview with an American newspaper early in 1998,[9] President Nobuhiko Kawamoto said Honda was willing to put "everything on the table" to fight global warming and reduce greenhouse gases. "As an industry, we should find some way to come together—not simply to talk, but to act—working together toward new solutions for our planet," he said. The offer is not unprecedented: In the 1970s, Honda let Ford look at its data on how to meet Clean Air Act goals without catalytic converters.

Honda's CVCC engine was an early low-emission entry, and the company is working on a methanol fuel-cell car it will show in 2005 or 2006, the fruit of work begun in 1989. The company's fuel-cell investment is increasing, but it's equally interested in cleaning up the gasoline engine, as shown by the Accord-sized Z-LEV motor it touts as nearly pollution-free.

Honda's California-based U.S. spokesman, Art Garner, is

noticeably pessimistic about alternative-fuel cars, an attitude reflected in the company's decision to cancel its EV Plus battery-car program after two years. "I agree with most people in the industry that fuel cells represent the best long-term potential, but they won't be here anytime soon," he told me. "It will be a long time before they're financially feasible. For the foreseeable future, gasoline power will be what counts." Could Honda be working with Ballard? "Everybody has talked to Ballard at one time or another," Garner replied.

Like Toyota, Honda is experimenting with both methanol and metal-hydride storage of hydrogen. It has multiple test cars, but rarely lets them out in public. I was, therefore, both surprised and pleased by Honda's invitation in the fall of 1999 to see and drive FCX-V1 (metal hydride) and FCX-V2 (methanol) at a Japanese race track.

As it turned out, only the Ballard-powered FCX-V1 was ready for driving duty. Its sister car, which Honda engineers exhibited with its fuel cell whirring and hissing, was deemed too noisy for press duty.

Both Honda fuel-cell test cars were built into the body of the discontinued EV Plus battery electric, though Honda exhibited an altogether different (and more aerodynamic) body style at the 1999 Tokyo Motor Show. Of the two cars, the not-ready-for-prime-time FCX-V2 is the more interesting, since it incorporates Honda's own fuel cell and reformer. Could the Japanese beat Ballard at its own game? And could the in-house engineers meet the significant technical challenge of miniaturizing an efficient methanol reformer?

It wasn't yet clear that they could. Both test cars had room only for a driver and passenger; fuel-cell equipment swallowed the back seats. And it soon became obvious that Honda would not be as free with its FCX cars as Ford was with the P2000. We were allowed only a single quarter-mile run around a parking lot, with a nervous company engineer in the passenger seat. FCX-V1 was certainly drivable, if somewhat slow and rattly. It felt like a car that had been in pieces the night before, and it may well have been. The desire to test fuel-cell cars under real-life conditions is one reason Honda decided to join Daimler-Chrysler in the California Fuel Cell Partnership.

I had wondered if Honda and Toyota were secretly sprinting past the competition from Germany, Canada, and America, but my brief test drive convinced me otherwise. In late 1999, this was still a close race, with no clear winner emerging.

OTHER REPORTS FROM THE GLOBAL VILLAGE

The power spots for the hybrid and fuel-cell car are the United States, Germany, and Japan, but that's not to say that nothing is happening elsewhere in the world. Other countries are making advances in cleaner car technology and commitment to environmental concerns, including the recycling of car-body parts. There's more to a green car than a clean power plant. By 1995, laws in many European countries had made it mandatory that 80 percent of automotive plastic be recycled, and most small plastic parts are marked for the recycling bin.

London's Fuel-Cell Cabs

I was startled to read, after hearing so many caveats about the slow pace of development, that London's first fleet of fuel-cell taxis went into action in the summer of 1998.[10] The ZEVCO Millennium doesn't give away its secrets easily: It looks like a standard London taxi, but under what the British call the "bonnet" it has an unusual alkaline fuel cell (most carmakers use PEM technology) whose purpose is to keep the battery array charged and supply the electric motor. The fuel cell, which runs on hydrogen gas stored under the cab's floor, is more like a range extender than a primary power supplier, and it's not the Holy Grail that carmakers are seeking, but it's quite interesting nonetheless.

The fuel-cell cab got a gala launch in front of London's Parliament building, complete with Greenpeace banners. Nick Abson, chief executive of the British-Belgian ZEVCO, proclaimed, "What we have developed is the world's first commercially viable alternative to diesel power which will set the standard for a cleaner Europe. It's a sea change in transportation technology."[11]

The cabbies seem supportive. "At first we were skeptical," said George Kay of the London Motor Cab Proprietors' Association. "But our drivers spend all day in the cab, and you're getting more pollution inside than outside the car." If state taxes can continue to support low-emission vehicles, "the cab trade will take to this new taxi like a duck to water," he said.[12]

The event was also attended by Chris Fay, the chief executive of Shell UK, a company with a dismal environmental record but a growing interest in hydrogen. Like the tobacco companies, which no longer want to link their future to selling cigarettes, oil companies like Shell are intent on expanding into other types of fuel.

Fay noted that Shell has established a Hydrogen Economy team dedicated to investigate opportunities in hydrogen manufacturing and new fuel-cell technologies in collaboration with others, including DaimlerChrysler. "We believe that hydrogen fuel-cell-powered cars are likely to make a major entrance into the vehicle market throughout Europe and the U.S. by 2005," he said, pledging that the British company was in the hydrogen business "for the long haul."

Fay's comments make an interesting contrast to the oft-stated opinion that fuel cells must run on gasoline, because the oil companies will block them otherwise. Both Shell and British Petroleum have pulled their U.S. operations out of the Global Climate Coalition, an anti-environmental business lobby. BP, now apparently a believer in global warming, has also withheld dues from the American Petroleum Institute in protest of its opposition to the Kyoto accords.[13]

Volvo's Ground-Up Approach

Sweden's Volvo, which was purchased by Ford for $6.5 billion in early 1999, has made strong environmental commitments. The company goes to extraordinary lengths to ensure that it uses recyclable materials in its cars and that its junkers actually get recycled. Some 75 percent of current vehicles do get reclaimed, but Volvo thought it could do better. In 1994, it became a partner in a new venture called ECRIS, whose goal was to keep the entire car out of the scrap heap. I visited the Jönköping, Sweden, headquarters of ECRIS soon after it was founded, and saw workers carefully dismantling old Volvos (plus the odd Saab) and loading the parts into bins for plastic, rubber, fabric, and glass. The company has pledged to reduce toxic materials in its manufacturing processes wherever possible and holds all its operations to high environmental standards.

Volvo has also investigated a wide range of alternative-powered cars and sells natural gas versions of some of its cars in Europe, Japan, and Australia. There are no plans to import the cars to the United States, because we lack a refueling infrastructure for natural gas, though in 1997 American journalists had a fine time driving a fleet of them to the Biosphere in Arizona, where they were wined, dined, and given a tour of the facility's distinct ecosystems.

Volvo supplied some components to a Renault Laguna station wagon with a 30-kilowatt fuel cell, running on liquid hydrogen. The Fuel Cell Electric Vehicle for Efficiency and Range (FEVER) car, partly financed by the European Union, was unveiled in 1997 and has a 250-mile range. Like DaimlerChrysler's NECAR III, the FEVER runs, but there's something of a space problem—despite being a station wagon, the fuel-cell car has room only for its driver. Volvo is also a partner in another European venture, the Capri project, which is managed by Volkswagen. The fuel-cell VW Golf runs on methanol, and Volvo developed its compressor, power converter, and energy-management system. Both cars were innovative but now represent dated technology in the fast-paced fuel-cell world.

Characteristically, Volvo has its own approach to the greener car. In 1992, it designed and built an incredibly sleek (at least by Volvo standards) aluminum-bodied hybrid Environmental Concept Car (ECC) that served as a response to the California emission mandates. In addition to the recyclable plastic panels and water-based paints you'd expect to find in a Volvo, the ECC has a series hybrid drivetrain combining a diesel gas turbine working as a generator, a substantial battery pack, and an electric motor. The system is rather complicated, but the car achieves both decent performance and low emissions, plus a four-hundred-mile range. Volvo made no efforts to get what could be a very expensive car to build into production, but it showcased the car and its different innovations on the show circuit.

Volvo's Olle Boethius, speaking from Sweden shortly before Ford bought the company, said that the future looks "a bit hazy," though he's pretty sure an electric drivetrain will be an important part of the mix. Volvo has its own fuel-cell lab in Gothenburg, where the company is based, and it was testing 50-kilowatt stacks bought from Ballard even before the Ford purchase. Still, says Boethius, "We would say 'no' to fuel cells today. DaimlerChrysler talks about forty thousand cars by 2004, and that's very ambitious, but how much will they cost? We don't, however, see anything that would make fuel cells impossible."

As it joins Ford in the Alliance, Volvo is not likely to suddenly become a fuel-cell leader, though its sturdy sedans might make good fuel-cell test beds. Hans-Olov Olsson, Volvo's U.S. president, said in an interview that "joining the Ford family will give us an opportunity to participate in the fuel-cell alliance; it's one of the benefits of the purchase." Volvo will bring to the group a holistic approach to automotive recycling and "life-cycle" environmental impact (from the plant to the junkyard) that could strongly affect the manufacturing process for fuel-cell cars.

Volvo's focus has been on bifuel natural-gas and gasoline cars and hybrids. By 1998 it was selling five hundred bifuel sedans a year, many of them to natural-gas utilities in Europe. In 1999, it announced that it was working on its own "power-split" hybrid, which automatically shifts from its electric motor to its internal-combustion engine, possibly for the U.S. market. The company's planning, Boethius says, is driven by the California legislation. "That was the trigger," he added, "and it is a good trigger." If any company is going to readily accept being told to produce environmentally sound cars, it's Volvo.

France's City Cars

Not all the EV news from Europe is about hardware. The French seem to be doing the most serious thinking about how best to use EVs in the urban environment, and EV rentals are more widely available there than anywhere else in the world. In a program that lasted from 1993 to 1994, PSA, which incorporates Peugeot and Citroën, rented fifty EV minicars to residents of La Rochelle, France. The residents, who ranged from city employees to bakers, chiropractors, and retirees, were asked to make exhaustive notes about where they went, what they did, and how the electric Peugeot 106s and Citroën AXs performed. "They covered nearly 300,000 miles," reported PSA executive vice president Jean-Yves Helmer at the 1994 EVS 12 in Anaheim, California. "And their first surprise was that they were driving a real city car with very good acceleration." Helmer reported mostly good news from his company's experiment (one driver said, "The EV changed my life"), but he cautioned that electrics will only have a future if "we can create a market and generate enough volume demand to make the vehicle competitive."[14] Later, in 1998, PSA continued its demographic experiments in Coventry, England, with sixteen more Peugeot 106s. It also tested the Touareg, an intriguing hybrid sport-utility, but made no plans to produce it.

To be fair, not all European EV experiments had results as satisfying as those in France. In 1994, German carmakers Daimler-Benz, BMW, Volkswagen, and Opel lent some of their electrics to regular folks who live on the northern island of Rügen. According to *The Wall Street Journal,* the residents had trouble remembering the limitations of their range, and many were unhappily stranded.[15]

The miniature EVs made by PSA are hardly distinguishable from the gasoline microcars plying the European streets. With $4-a-gallon gas, it's not surprising that European drivers will readily accept tiny electrics.

Getting them into the Greenpeace Twingo might be even harder. In 1995, the European arm of Greenpeace moved from auto antagonist to auto designer with its own experimental version of the green car, a much-modified Renault Twingo. Like hanging a banner on a corporate skyscraper, Greenpeace's energy-efficient Twingo, which was never seriously considered for production, called attention not only to the world's auto addiction but also to the tendency of new models to gain weight and lose fuel economy.

Greenpeace's car, stripped of unnecessary poundage, runs on an exotic, two-cylinder supercharged gasoline engine and gets as much as seventy-eight miles a gallon. The car is not a hybrid, and does produce nitrogen oxides and carbon dioxide, but it's certainly good on gas. Greenpeace's Wolfgang Lohbeck can perhaps be forgiven a little arrogance when he said, "We are not car manufacturers, but there is so much incompetence among senior engineers that we had no choice but to show the car industry how to do it."[16]

It's heartening to see that the automobile's environmental reform is not an isolated effort but truly a worldwide concern. If EVs are going to take root in the United States, we'll have to learn some lessons from the Germans about long-term research programs, from the Japanese about federal support for alternative energy, from the Swedes about holistic environmental programs, and from the French about adapting the EV to urban life. And maybe Greenpeace can add a few words about eliminating weight to achieve fuel efficiency.

CHAPTER 7

THINKING ABOUT TOMORROW: VISIONARIES, PESSIMISTS, AND INVESTORS AT THE CROSSROADS

I NOW HAD A fuller picture of EV progress around the world, but no clear idea about how it would all fit together. Slowly, the global auto industry was committing itself to profound change, but it was not altogether certain how the new products in the pipeline would be received by consumers. And what about the infrastructure? Battery EVs require charging stations, not only at home but on the road and in workplaces, too. Fuel-cell cars, if they're not to be run on gasoline, require a vast, multibillion-dollar hydrogen or methanol refueling network to replace the corner gas station. America's hydrogen-generating capacity is now minute, and only California has any kind of statewide EV charging system. I needed a crystal ball to gaze into this uncertain future.

Fortunately, there's no lack of vision on alternative energy, though those who are focused on the future see different things. I talked to both optimists and pessimists about how the clean-car scenario would play out in the next decade and beyond.

AMORY LOVINS AND THE ROCKY MOUNTAIN INSTITUTE

Amory Lovins is co-CEO and co-founder of the Rocky Mountain Institute (RMI), a diverse think tank founded in Snowmass, Colorado, in 1982 that works with industry to pursue what it calls "soft" or sustainable energy paths. In a highly prescient 1995 article in *The Atlantic Monthly,* Lovins sketched out the auto industry's future before the industry itself knew what was coming. Writing with his wife and partner L. Hunter Lovins, he predicted, "Well before 2003, competition, not government mandates, may bring to market cars efficient enough to carry a family coast to coast on one tank of fuel, more safely and comfortably than they can travel now, and more cleanly than they would with a battery-electric car plus the power plants needed to recharge it."[1]

The Lovinses were talking about what they call "hypercars," ultralight and superaerodynamic vehicles made of advanced polymer composite material and powered by efficient hybrid electric drivetrains. Battery electrics, they noted derisively as GM was struggling to bring out its EV1, have to "haul a half ton of batteries down to the store to buy a six-pack."

The cars that will come to market in the Lovinses' timeframe aren't quite hypercars, and they won't travel across the country on a single fill-up. But when Ford, for instance, brings out hybrid and fuel-cell versions of its lightweight P2000, it will have gone a long way toward fulfilling that vision. And seeing that it would happen from the vantage point of 1993 (when he actually wrote the *Atlantic* article) is truly remarkable.

As a unit of RMI, the Hypercar Center was founded in 1993, and ultralight hybrids and fuel cells have become a major focus of the group. Lovins works closely with senior research associate Brett Williams, who said the center has gone from "selling the idea" to "talking about how we think it should be done."

Amory Lovins started out as a prodigy, pursuing graduate work in physics at Harvard while still a teenager. He joined the faculty of Oxford's Merton College as a junior research fellow at twenty-one. From the beginning, he studied American energy policy and concluded that our reliance on coal and nuclear power was wrongheaded. The United States, he said, could operate on a quarter of the energy it consumed if it would just start to use the more efficient motors, lighting, building construction, and production methods that were already available. Instead of just talking about it, he founded RMI with his wife in 1982, and the nonprofit organization serves as a consultant to many large industries (including car companies).

Lovins himself has been an energy policy adviser to more than thirty countries, has briefed ten heads of state, and has written twenty-six books. He has become an energy eminence, and I met no one in the car industry who didn't at least know his name. Lovins "doesn't have all the baggage that Detroit has," said one Big Three staffer. "He doesn't say, 'We've always done it this way.' " John Wallace of Ford calls him "Amory 'No Job Is Too Hard for Someone Else to Do' Lovins." And he seems to mean that affectionately. It's not surprising that Lovins's relationship with the auto business is somewhat love-hate, in that he is severely critical of its current practices.

Noting, for instance, that 80 to 85 percent of a car's fuel energy is lost before it gets to the wheels, Lovins chides "the appalling waste" of building cars from heavy steel and estimates that the industry could save 65 to 75 percent of the weight by building cars from advanced materials, many of which are actually stronger than (and offer safety advantages over) steel. GM, he says, demonstrated that it could do that with its 1991 Ultralite prototype, which weighed only fourteen hundred pounds but could haul four passengers at 135 miles per hour and could average sixty-two miles per gallon, with a New York-to-Los Angeles range. That car was powered solely by a direct-injection three-cylinder engine, but just think what it could do with a hybrid drive![2] Lovins thinks two hundred miles per gallon is possible.

The auto industry sometimes publicly criticizes Lovins for not understanding the "real world" of car production. They point out, for instance, that lightweight aluminum costs four times as much per pound as steel. But since Lovins first started talking about composites, the materials have become common in production cars, and are almost universally contemplated for hybrid and fuel-cell designs. Yes, composites like carbon fiber are also much more expensive than steel per pound, but Lovins predicted early on that the costs would come down dramatically with mass production, and that has indeed happened.

Of all the pundits I talked to, Lovins is probably the most optimistic about the fuel-cell time line. He's also, as a consultant to the industry, the environmental authority with the most insider knowledge. So when he says that "there are good reasons to expect to see a Toyota fuel-cell car on at least the Japanese market before 2002," and that the auto companies committed $5 billion to secret "black budget" ultralight hybrid research by 1998, it's worth paying close attention, even if the auto industry laughs off these contentions. And a hydrogen-energy economy? Well, that's a given. "I think the odds are close to 100 percent," he says.

In a lengthy interview, Lovins outlined how he thinks we'll get to that hydrogen economy. It will start with "distributed" power, stationary fuel cells that generate on-site power in apartment buildings or hospitals. Lovins assumes that the manufacture of fuel cells, rather than their hand-assembly by Ph.D.s, will bring the cost down from $3,000 per kilowatt to $800. "An $800 to $1,000 per kilowatt fuel cell makes perfect economic sense in a building," Lovins says. "The space and weight don't much matter when it's sitting in a basement or outside the building. An additional benefit is that you can use the waste heat the fuel cell generates through a co-generation process to provide building services like heating, cooling, and dehumidification. Now instead of the fifty percent efficiency of a fuel cell with a reformer, or sixty to seventy percent without one, you can actually get into the nineties of total system efficiency. It turns out that in most situations, the waste heat is worth enough as a commodity to pay for your natural-gas line and a mass-produced reformer to turn it into hydrogen. And once you've done that, the effective net cost of delivering electricity to your building is on the order of a cent or two per kilowatt hour. Right now, commercial buildings pay an average of six cents. So I think the building market comes first, then you get the prices down and put fuel cells in cars."

Once the stationary fuel cell catches on, costs should drop quickly. "The building market is rather large," Lovins says. "Buildings use two-thirds of all the electricity in the United States, so you could build very large fuel-cell production volumes. Actually, both the building and vehicular fuel-cell markets are potentially so big that when either of them starts to happen, it makes the other one happen, too, by building volume and cutting costs. It could go either way."

Stationary and mobile fuel cells could have a symbiotic relationship that goes beyond cost, Lovins believes. "Once you put a fuel cell in an ultralight car, then you have a twenty- to twenty-five-kilowatt power station on wheels, which is driven about four percent of the time and parked ninety-six percent of the time. So why not lease those fuel-cell cars to people who work in or near buildings where you've already installed fuel cells?"

It would work like this: Commuters would drive their cars to work, then plug them into the hydrogen line coming out of the reformer installed as part of the building's fuel cell. While they worked, their cars would be producing electricity, which they could then sell back to the grid at a time of peak power demand. The car, instead of simply occupying space, would become a profit center. "It does not take many people doing this to put the rest of the coal and nuclear plants out of business," says Lovins, who's been trying to do just that for decades. "The hypercar fleet will eventually have five to ten times the generating capacity of the national grid."

Thinking about cars as power plants is not something that Americans are conditioned to do, but it's a good indication of why fuel cells change all the rules. With conventional wisdom, auto industry insiders say building a hydrogen infrastructure in the United States will cost hundreds of billions, since there is a very limited hydrogen-generating capacity now. Decentralizing production, by putting reformers in buildings and even in home garages in conjunction with local power generation, reduces that prohibitive cost. That said, larger reformers in small neighborhood facilities are the "gas stations" of tomorrow.

Lovins is full of visionary ideas for knocking the bugaboo known as "hydrogen infrastructure" down to size. He envisions older hydroelectric dams being converted to "hydro-gen" dams, a transition that makes economic sense, he said, because hydrogen delivers three or four times more energy at the wheels of a fuel-cell car than does the same amount of gasoline in today's cars. "The extra efficiency in the hydrogen justifies producing it at the dam, shipping it through a pipeline, and distributing it," Lovins says. "The hydrogen equivalent of $1.25-a-gallon gasoline is nine to twelve-cents-a-kilowatt electricity. The people who run hydroelectric dams would love to sell their energy for nine to twelve cents because all they can get in the Northwest market is 1.6 cents and falling. With more fuel cells around, the electricity prices will fall even more. Dam owners can make enormously more profit selling hydrogen than selling electricity."

Lovins is that rarity in the environmental movement, an inherent optimist who believes that technology will eventually overcome some of our most pressing problems. He quotes Israeli diplomat Abba Eban, who said that people and nations behave wisely once they've exhausted all the other alternatives. A case in point is the industry's fixation on the methanol fuel-cell car with an onboard reformer. "Our conclusion is if you have a good reformer, it would work a lot better out of the car," Lovins says. "There's no reason to have the reformer in the car, unless you feel that the car has to run on liquid fuel."

Lovins may be a bit quick to dismiss some of the very real problems automotive fuel cells need to overcome, but he's brilliant at envisioning how they could fit neatly into a clean-energy economy based on renewable hydrogen. That may seem like an impossible dream but, five years ago, so was the idea that the auto industry would bring fuel-cell and hybrid cars to market.

VOICES FOR FUEL CELLS

If RMI was a sole voice calling for direct hydrogen in fuel-cell cars, it would have a hard time getting heard. But the institute's position is bolstered by respected figures such as Ford consultant Sandy Thomas, vice president of Directed Technologies, a high-technology consulting firm. Thomas concluded in a research paper presented to the 1997 U.S. Hydrogen Meeting that investors could make a 17 percent financial return by building a refueling network for fuel-cell cars.[3] "People say it would cost $60 to $70 billion just to get hydrogen into pipelines," Thomas said in an interview. "We're saying, don't do that. Make the hydrogen where people want it: at the filling stations, at fleet operators' garages, and even at home. Through economies of mass production, that could be the least costly way to make hydrogen in the long run."

Joan Ogden has come to the same conclusion. A Princeton research scientist who runs the university's Center for Energy and Environmental Studies, Ogden has been a tireless voice at conferences championing the cause of direct hydrogen. In a recent paper that extensively studied the near-term hydrogen capacity of the Los Angeles region, she asserted that "hydrogen infrastructure development may not be as severe a technical and economic problem as is often stated. The hydrogen fuel option should be considered viable for fuel cell vehicles, and development of hydrogen refueling systems should be undertaken in parallel with fuel cell vehicle demonstrations."[4]

Ogden, one of the few women in the thick of this debate, said in an interview that she began studying hydrogen fuel cells ten years ago, prompted by the urgent need to reduce global warming and control what would otherwise be the massive spread of pollution-spewing tailpipes to the developing world. "I came to this looking at the technical feasibility," she says. "Fuel cells offered a quantum leap in terms of improved efficiency and reduced emissions. But did we have the available resources to make direct hydrogen work? I became convinced that we did. In the near term, it looked interesting and doable. At the center, we've put together a computer model of the fuel-cell car and developed a series of equations for how much power you'd need at the wheels, given the weight of the car, the energy density of the fuel, the accessories you'd need, and so on. With an onboard reformer, you could reach seventy miles per gallon, but with compressed hydrogen you could achieve the equivalent of one hundred miles per gallon." These figures contrasted to the sixty to seventy miles per gallon achievable in hybrid cars with small direct-injection diesel engines, which have much higher emissions.

Ogden admits that direct hydrogen refueling entails added infrastructure costs, but these costs are roughly comparable per vehicle to the expense of onboard reforming, which adds weight and robs space. "I've talked to the industrial producers, and it need not be unthinkably expensive to build a hydrogen-production infrastructure," she says.

Like Lovins, Ogden thinks we are moving to a hydrogen economy—eventually. "I don't see us all driving around in internal-combustion sport utilities forever," she says. "The best future scenario is a fuel-cell car carrying hydrogen gas, but how we get to that, I'm not sure. We may go through a stage with methanol." It's possible, Ogden says, that we'll enter a period like that between 1900 and 1910, and no time since, when a variety of automotive fuels competed for dominance. "We went from whale oil to electricity," she said. "Now maybe we can go from gasoline to hydrogen."

I was learning, in theory, that a totally clean energy loop was possible. But was there anyone actually in the trenches trying to close that loop? Some of the most promising work is being done at California's Humboldt State University, which has developed its own running fuel-cell car, the first of its kind in America, and is designing a refueling network for it, too. The Humboldt car, a converted Kewet El-Jet like the noisy fiberglass box I drove in North Hollywood, was built by the faculty at the college's Schatz Energy Research Center. In 1998, it joined an existing fleet of three surrey-roofed fuel-cell golf carts the school built as part of a pilot project under way in the retirement community of Palm Desert, California. All four vehicles are in regular use for grounds maintenance and commuting.

There's more to Humboldt State's fuel-cell project than some interesting engineering challenges. The college is in Arcata, one of the most environmentally enlightened communities in the United States, with its town government controlled by the Green Party. Hemp is practically a religion in Arcata, which also has an enormous food co-op, a pond that filters sewage, and meditation redwoods. Arcata is, needless to say, a nuclear-free zone.

Ron Reid, a research engineer at the Schatz Center, says everybody in Arcata is "pretty proud" of the fuel-cell car, which, strictly speaking, is a hybrid that runs mainly on batteries, like the ZEVCO London taxis. The car "tanks up" at a temporary hydrogen station the school has set up in Palm Desert, but under construction is an elaborate refueling operation, designed to be both user-friendly and automated, that will make its own hydrogen on site. What's more, the hydrogen will be made from solar energy, a completely clean system from the photovoltaic cell to consumption in the car. The Schatz engineers estimate that with a 340-square-mile area of desert, they could produce enough local photovoltaic hydrogen to power fuel-cell cars for the whole city of Los Angeles—an amount equivalent to the fifteen million gallons of gasoline existing city cars use daily. Decentralizing production means a whole chunk of that much-feared infrastructure—specially prepared tanker trucks and leak-proof pipelines—won't have to be built.

"Our ambition is to have renewably produced hydrogen be the fuel of choice in our society," says Reid, who spends a lot of time educating students on the possibilities of the world's lightest gas. Like Lovins, Reid is a true believer in hydrogen's power to deliver us from fossil fuels, and the university's project in Palm Desert is a landmark in turning that environmentalist's dream into practical reality.

READ ALL ABOUT IT: AMERICA'S NAY-SAYING AUTO PRESS

You would think that America's auto-enthusiast press, characterized by monthly magazines such as *Road and Track* and *Car and Driver* (which are jointly owned), *Automobile,* and *Motor Trend,* as well as more frequent periodicals, including *Auto*

Week, would have been on top of fuel-cell and hybrid-car developments. You would think that, but you'd be wrong. No visionaries here. Enthralled with the horsepower race, beholden to carmakers for advertising, the auto press is brilliant at presenting the thrill of fast driving and the excitement of new $100,000 Ferraris, but it mostly ignores alternative fuels and can be downright hostile to environmental initiatives. (An exception is *Popular Science,* whose Dan McCosh has taken a personal interest in EVs. The Prius was the magazine's 1997 "Best of the New" car winner.)

Automobile, for instance, features the liberals' scourge, P. J. O'Rourke, who laps up statistics from anti-environmental think tanks such as the Competitive Enterprise Institute and the American Enterprise Institute. In one piece, his "smoking gun" is the assertion that cars and trucks contribute only 1.5 percent of global greenhouse gas emissions. "Wouldn't the sputter budgets be better off applying their efforts elsewhere?" he asks in a column adjacent to an ad for eighteen-inch sport-utility wheels. "Maybe they could limit the amount of hot air released by stuffing their socks in their mouths."[5]

O'Rourke *is* funny, but the statistic is ludicrous. In the United States, cars and trucks contribute 25 percent of the carbon dioxide emissions that cause global warming. (Around the world, where cars aren't quite so prominent, the figure is 14 percent.) Every year, the average American car pumps five tons of carbon dioxide into the atmosphere.[6]

Writing in *Car and Driver,* veteran auto journalist Brock Yates took aim at the usual environmental targets, including "Ozone Al [Gore]," but reserved his particular ire for fuel cells, invoking the *Hindenburg* in the process. The fuel-cell car, he said, had "the makings of another technological dead end now occupied by the dismally failed GM EV1 and other all-electric farces." Even worse, he said, the car's exhaust note sounded like something produced by "barnyard animals."[7] Someone should inform Mr. Yates that fuel-cell cars don't really have exhaust notes, barnyard or not.

Automobile's editor, the highly respected industry veteran David E. Davis, admits he doesn't know much about fuel-cell and hybrid cars. "I still don't understand the fuel cell," he volunteered cheerfully in an interview. "I should know more about that, but I'm completely innocent of any knowledge."

Davis, who assails Ralph Nader in his magazine column ("that sucker is *dead,*" he wrote) and attacks government regulators as "safety Nazis," nonetheless seems open to the idea that the internal-combustion engine may be in its last days. "Everything is replaceable," he said. "It does such a brilliant job so cheaply, but I can't imagine it won't be retired sooner or later. We've certainly managed to replace reciprocating engines in airplanes."

And Davis claims that new technologies are good for the automotive press. "Innovations are what keep magazines alive," he said. "Listen, the important thing for most car enthusiasts is good dynamics and a good power-to-weight ratio. If you can have that in a fuel-cell car, powered by hydrogen or body odor, great."

Car and Driver editor-in-chief Csaba Csere, a former Ford engineer, pays more attention to the new technologies, but he doesn't think Americans will buy them. "The weakest market in the United States is small cars. Saturns are hard sells, Cavaliers are hard sells. The luxury sedan is one of few segments that is actually growing. The Prius has less utility than a $13,000 Ford Escort, and it's priced thirty to fifty percent higher than equivalent small cars, with less performance and a little more hassle. How many consumers are so concerned about global warming that they'll spend more money for an inferior automobile? Look at the sales of the EV1—they're a drop in the bucket."

Like most auto journalists, including Davis, Csere thinks the bottom line is what consumers *want,* and social responsibility doesn't enter into it. Besides, he's not convinced that global warming is a real problem—though, interestingly enough, Csere does say that if global warming *were* proven, he would support an across-the-board carbon tax increase.

And though Csere admits that "the fuel cell seems like the best solution," he dismisses it as an unproven technology that still costs too much. Why hasn't *Car and Driver* covered the cell's dramatic progress from dreamer's fantasy to international competition? "We don't tend to cover developments when they're still out there," Csere said. "And nobody expects a fuel-cell car sooner than ten years from now. Where would you refuel it? I don't know of any gas stations offering liquid hydrogen or methanol. Hydrogen gas is fairly volatile, and you can't put out a hydrogen fire with water. These infrastructure problems take time. Auto companies are in the business of providing personal transportation, and gas engines, which are getting cleaner all the time, are still the best means of providing it." Still, looking ahead to 2025, Csere said, "We probably will be driving fuel-cell cars; they'll maybe be ten percent of the total."

Veteran auto writer Robert Cumberford, now based in Paris, says candidly that car magazines don't write about clean cars because there's not much to get excited about. "Remember that a lot of automotive enthusiasm is based on what is undoubtedly immature excitement over excess," he said to me in a 1998 e-mail. "Excess power, excess speed, the ability to smoke the tires under extreme acceleration, and so on. A silent system that simply gets the job done is not exciting, and thus not newsworthy. . . . [And] for many people there is a strong emotional attachment to cars—there are no books or magazines dedicated to refrigerators."[8]

But colorful visions into the next century do make good reading. A glance through the archives shows that the magazines used to look into the future far more than they do now. While fuel cells may not be just around the bend, hybrid cars are, and the magazines' readers might enjoy reading about them. In its coverage of the 1997 Tokyo Motor Show, *Automobile* took far more interest in Daimler-Benz's twelve-cylinder Maybach luxury car and a quartet of showgirls in skimpy Mitsubishi jerseys than it did in the modest Toyota Prius, which got only a passing nod (and no pictures). "In all, the Japanese soberly marched to the show's theme: 'Harmonize vehicles to society and environment,' " the magazine wrote. "But excess is always more fun than sobriety, so it's no surprise that the star of the show came from the other side of the aisle."[9] The car in question was a 440-horsepower BMW sports model. Turn the page and you came to a two-page ad for the Infiniti QX4 sport-utility, shown fording a stream. "Careful," the ad warned. "You might run out of planet." What's more likely is that sport-utility vehicles like that one will run the planet we have into the ground, with the car magazines cheering them on.

FUTURIST PREDICTIONS

The fuel cell, with its promise of zero-emission cars wafting along, dripping drinkable water out of their "exhaust" pipes, is a favorite of futurists, many of whom paint a remarkably benign portrait of our technological liberation in the twenty-first century.

Peter Schwartz, for instance, chairman of the San Francisco–based Global Business Network (GBN), looks past the stock market scares and Asian defaults of 1998 and sees the world in the beginnings of a "long boom," aided by global telecommunications hookups (video phones in developing countries!) and biotechnology breakthroughs.[10] It's the kind of vision beloved by *Wired* magazine, and it's not surprising that the bearded guru's ideas were featured in a 1997 cover story there. Stewart Brand, founder of *The Whole Earth Catalog* and the Well, a futurists' computer network, is a GBN co-founder.

Although it's been wrong on occasion, as when it missed the looming Mexican financial crisis in 1994, GBN is more often right. An international business with clients such as IBM, AT&T, and the Pentagon, GBN is also heavily involved in the auto industry. It was instrumental in introducing Daimler-Benz to Ballard Power Systems in 1996, and the company has also worked with GM, Ford, Volvo, Nissan, Toyota, Honda, and Fiat.

"Something quite dramatic happened in the auto industry [in 1998]," Schwartz said. "There's been a wholesale flip, a sea change, across the whole auto industry from profound skepticism, doubts, and halfway commitments to an almost inevitable move towards fuel cells and EVs. All the companies are in one way or another pursuing this."

Schwartz predicts that by 2020 the hydrogen fuel cell will have won the war with internal combustion. "We'll see the first cars on the road in early 2003 or 2004, and by 2010 the majority of the cars in the showrooms will be electric drive, including battery EVs, hybrids with direct-injection diesels, turbo generators and fuel cells," he said.

Schwartz—and Lovins, too, for that matter—doesn't think that the move to fuel cells will be prompted by declining oil stocks. "The cost of developing new oil discoveries continues to fall, and it's hard to see a drop in productivity," he says. "There was supposed to be 1.5 billion barrels of oil in the North Sea, but now there appears to be six billion barrels. I don't think we'll hit the physical limits of oil production until 2040 or 2060."

What, then, will drive the shift to fuel cells? "The external driver is climate change," Schwartz says. "Global warming is bigger than most people perceive, and more urgent. All the auto companies are convinced it will be a big deal." Beyond that, EVs are just, well, cool. "I've driven them and they're fast, hip, and modern," says the futurist. "People will look at the old, smoky, gas-burning mechanical technology and see it as incredibly dated. There's absolutely no reason young people won't buy them, because they'll be like sport-utility vehicles are today."

Another futurist with interesting things to say about the auto industry is the Arlington Institute's John Petersen, who Schwartz calls an "out-of-the-box thinker." In Petersen's latest book, *Out of the Blue,* he speculates what will happen if air travel moves at faster-than-light speeds, if self-aware machines straight out of *The Terminator* are developed, and if human cloning becomes legal. And he also looks at the end of the carbon economy and the replacement of internal combustion with fuel cells.

There's no doubt that Petersen is a rosy optimist. He speaks glowingly of "a new generation of kids growing up around the planet who are very sensitive to the environment, but at the same time are on top of technology." By 2025, he thinks, these new thinkers will be aided by tiny chips that will have fifty times the computing power of the largest three-year-old supercomputer, "yet they'll only cost a penny to manufacture."

With that staggering vision, it's not surprising that Petersen says "the writing's on the wall" for internal combustion. "It's clear to all the manufacturers that they will be out of the gasoline engine business," he asserts. Lighter, stronger structures will make electrical drives feasible, he adds, with the impetus to develop that technology coming from legislative mandates and an increased sense of competition. "If we can get China to embrace the fuel cell, not only in terms of automobiles but for electrical power, they can leap across the industrial age as we have traditionally defined it," says Petersen, who has given lectures to audiences as diverse as the Wharton School of Business and the Joint Military Intelligence College.

Futurists have been wrong before. They did, after all, predict that nuclear power would produce electricity that was "too cheap to meter." In direct contrast to the neo-Luddites, they tend to think that new technology will free the human race from its current woes and solve a host of environmental problems. Most valuably, they capture the sense of excitement and innovation that can develop around new inventions, from television to the Internet. If fuel-cell cars are introduced, they seem to be saying, won't human ingenuity find a way to make them work? And given that likelihood, shouldn't we be investing in them now?

FINANCIAL ANALYSTS ARE BULLISH ON FUEL CELLS

Fuel-cell development has mostly flown beneath the radar screen of public recognition. But for those investors lucky enough to recognize what was happening, fuel cells have already proven extremely lucrative.

Without actually selling any mass-produced fuel cells or even making a profit, the publicly traded stock of Ballard Power Systems has soared. On the Canadian Stock Exchange, it reached the giddy heights of $182 (Canadian) a share in March 1998 before settling back. Analyst Robert Chewning of Morgan Stanley Dean Witter in New York said the runaway success of the stock at that time could be attributed to the company adding GM, which had previously gone it alone, as a fuel-cell customer. "The fact that [GM is] coming to Ballard and ordering a fuel cell—what do people read into that?" Chewning asked. "Obviously, this has attracted a lot of the media."[11]

On the NASDAQ, the stock climbed from an initial price of $10 a share to a high of $60, allowing some savvy early investors (who also benefited from a three-for-one stock split) to walk away with eighteen times their outlay. Robert W. Shaw Jr.'s Arete Corporation, a New Hampshire–based venture capitalist firm with $135 million in assets, was an early investor in Ballard, starting out in the late 1980s when meetings with the company were "five people around a table." Ballard was then, he says, "a $100 million company without a lot of support from the U.S. and Canadian governments." Now no longer invested in Ballard, Shaw thinks that if the company "doesn't deliver something soon—a real, ready-to-be-manufactured product," its stock could suffer. Tangible results are definitely reflected in the stock price. In April 1999, when Ballard announced a California fuel-cell test with DaimlerChrysler, its stock jumped 11 percent.

Shaw is almost as much a true believer in fuel cells as Amory Lovins and shares Lovins's conviction that a hydrogen-based energy economy is almost a certainty. Shaw spends his time surveying a vast field of new start-ups and spinoff ventures, trying to find what he calls "the next Ballard."

Another company on the hunt for novel energy ventures is Nth Power Technologies, which since 1996 has raised $65 million for such investments. Nancy Floyd, co-founder of the San Francisco–based company, thinks the era of what she calls "highly inefficient, cathedral-style central power plants" is over and in its place will be small-scale distributed energy, with on-site generation and storage. And that's why her company is seriously investigating photovoltaic and fuel cells, which are local power generators. "The fact that they're environmentally benign is an extra," said Floyd, who earlier founded a company that successfully commercialized wind energy. In noting the demise of companies like flywheel pioneer Rosen Motors, however, Floyd admitted there are dangers when companies serve as parts vendors to an automotive industry that may or not want what they're selling.

Investment in many forms of alternative energy skyrocketed in the late 1990s, said Jeanne Trombly, a consultant to San Francisco–based Investors' Circle, whose members commit venture capital to a wide variety of socially responsible companies. "These new technologies seem to be breaking through the gate," she said, "in response to global warming and tightening federal and state regulations. The great thing is that, just when alternative-energy investment starts to make sense, there's a massive technological acceleration."

RALPH NADER: A PESSIMIST WITH HISTORICAL PRECEDENT

It's impressive that so much conviction about the future of fuel-cell and hybrid cars is being backed up with investment and research commitment, but does that mean the clean technologies will really be served up by a smiling, suddenly energy-conscious auto industry decked out in green plumage? After all, the carmakers had to be dragged screaming into production of such seemingly innocuous innovations as seat belts, catalytic converters, and PCV valves. Talking about his company's tooth-and-nail opposition to the Clean Air Act of 1970, GM spokesman Greg Terry recently admitted, "In a way, we made liars out of ourselves because we sincerely believed we couldn't pull a rabbit out of a hat."[12] Former EPA official Michael Walsh says the industry's "public posture is always much more pessimistic than technical reality. When you give them a challenge, they meet it."[13]

The 2003 California emission mandates are certainly a challenge, and the automakers are meeting them with a combination of battery EVs, hybrids, and fuel-cell cars. But will they go further than the law demands and actually replace the internal-combustion automobile, a vehicle that carries with it ultralucrative servicing and parts businesses? Are they willing to sacrifice their symbiotic relationship with Big Oil on the altar of public interest?

Ralph Nader would say no. The author of *Unsafe at Any Speed,* who wrote back in 1965 that carmakers would fight to the death against safety and pollution laws even if it meant the certainty of turning customers into victims, hasn't mellowed any. In fact, he recently updated the seminal book and comes down just as hard on the carmakers as he did back then.

In 1994, I attended a seminar at the New York International Auto Show entitled "How Will the Car Meet the Environmental Challenges of Tomorrow?" The speakers were EV advocate Ed Begley Jr., former transportation secretary and president of the now-defunct American Automobile Manufacturers Association Andrew Card—and Ralph Nader. While Card, who now works at GM, predictably spoke in favor of letting competition handle the problem, and Begley touted the coming wave of electric cars, Nader was decidedly pessimistic. He predicted that cars *would not* meet the environmental challenges, because California's fledgling attempt to regulate the industry would fail. The regulators would cave in to the lobbyists, he said, and the 1998 California mandate would be aborted, leading to the inevitable withdrawal of similar legislation in New York, Massachusetts, and other eastern states. He was right. By consistently betting that the desire for short-term profit would carry the day, and that government functions as an arm of industry, he's been proven correct time and again.

"The auto industry is in a period of functional stagnation," Nader said in a 1998 interview from Washington. "And I don't see anything in the next five years that will disturb that stagnation. Sport-utility vehicles are a nightmare that produces more pollution and less efficiency. The National Highway Traffic Safety Administration is an arm of industry, whose head doesn't know he's supposed to be a law-enforcement official."

When he was five years old, the Winsted, Connecticut, native was taken to the 1939 World's Fair, where General Motors showed off its streamlined wonders. "They knew how to dangle their innovative products in front of you, but the company itself has never caught up with the pavilion it built sixty years ago," Nader said. "They play a game with innovation, and it isn't meant to be serious. Essentially, they say, 'We have all these breakthroughs coming, so why regulate us?' It's a very astute maneuver by Detroit, which doesn't even want to build cars anymore—it makes more money on trucks. If the automakers had their way, the federal corporate average fuel economy (CAFE) mileage standards would be withdrawn, not tightened. As long as they exist, they can be extended."

In 1996, Nader said, he had lunch with Amory Lovins, who told him that the era of confrontation with the auto industry was over, that GM officials had responded favorably to his hypercar ideas. Nader was unmoved. "I said, 'Amory, everyone has to learn their GM lesson. The guys at the top may give lip service to environmental issues, but they're not serious. To make it happen, there has to be a whole new consumer movement pushing for it. Last month's receipts are more important than the next decade's impact on the environment.' "

Ask Nader about the looming problem of global warming, and he'll say worrying about it is just not on the corporate agenda. "What do the car executives care?" he says. "They'll be retired before it makes any difference." Nor does Nader think that declining oil reserves are a motivator. "You can go back and look at predictions that we'd be down to our last drop by 2000, but there's more oil in 'found' reserves now than there was in

1975. I think there is no end to oil even in the next century, and until we really do run out, there's no market incentive for cleaner cars."

One wants Nader to be wrong. Even *Nader* probably wants Nader to be wrong. Certainly, visionary leaders have come out of corporations. But out of GM? Nader doesn't even think the corporate culture would allow it to happen. In the 1986 book he wrote with William Taylor, *The Big Boys,* Nader details the contents of an uninhibited memo sent to senior management by departing Chevrolet general manager John Z. DeLorean in 1972. DeLorean had just been told by GM's president that he was too concerned about social issues and lacked "single-minded focus and dedication to the immediate welfare of the corporation."[14]

Much of DeLorean's nineteen-page response was about safety issues, and he chided the corporation for not spending more, especially when only a few cents per car would have saved lives. But then he got to emissions. "In no instance, to my knowledge, has GM ever sold a car that was substantially more pollution-free than the law demanded—even when we had the technology," DeLorean wrote. "As a matter of fact, because the California laws were tougher, we sold 'cleaner' cars there and 'dirtier' cars throughout the rest of the nation. . . . Our corporation has lost credibility with the public and the government because each new emissions standard has been greeted by our management's immediate cries of 'impossible,' 'prohibitively expensive,' 'not economically responsible'—usually before we even know what is involved."[15]

You could dismiss DeLorean as an unreliable source, given his subsequent history with the car that bore his name, but his was for that brief moment an impassioned voice for social responsibility from the heart of America's biggest corporation. And because he didn't think that such concern for the greater good was possible at GM, he left the following year. Twenty years later, GM was still selling cleaner cars in California than it was in the rest of the country.

The business of America is business, and GM's bottom line corporate culture is hardly unique. Getting inside that culture is the single-minded obsession of Brad Snell, a friend of Nader's and a former assistant Senate counsel. Snell has been working on a muckracking history of General Motors since 1980, and he's equally convinced that profound change isn't likely to descend from the corporate suites in Renaissance Center. "They drive out the electric cars when any environmental concern is raised and say, 'See, we're working on it.' When the furor dies down, you never see them again," Snell says. "The economics aren't there. They can't make as much money from EVs because they have fewer moving parts. They don't wear out as quickly and there's not as much room for 'aftermarket' sales once the car is out of the showroom. If the fuel-cell car has more aftermarket, then it will be their fallback position. I don't perceive this as malevolent: It's a profit-maximizing strategy, and the internal-combustion car is their greatest profit center. If all the research that had gone into internal combustion had instead gone into EVs, you'd have fabulous batteries now."

It's hard to deny the relentless logic that people like Ralph Nader and Brad Snell bring to their analysis of automaker behavior. And as John DeLorean points out, automakers, particularly domestic ones, have seldom made safety or environmental improvements that weren't forced on them by regulation. No one's making Detroit produce hybrid or fuel-cell cars, though the California legislation and tightening federal emissions standards are a powerful driver.

But competition is a driver, too, and the Big Three could precipitously lose market share (as they did in the 1970s, when Americans started buying Toyota Corollas and Honda Civics in record numbers) if they're caught with nothing when an innovative new car appears from Germany, Japan, or even Korea. The Toyota Prius is a car like that, and it had a remarkable effect in stimulating hybrid development plans in the United States. If Amory Lovins is right, and Toyota brings out an early fuel-cell car too, well, that would be very embarrassing for Detroit.

GM is a cagey company. It rolled out a series of innovative hybrid and fuel-cell prototypes, and chairman Jack Smith made impressive statements about the company's direction. But unlike DaimlerChrysler, which vows to actually start making fuel-cell cars in 2004, GM made no production commitments and hedged its bets. The company's wide-ranging deal with Toyota, however, sends a strong signal that it will, in the end, proceed to market.

There's certainly no unanimity of views about clean cars. For every visionary, there's a naysayer. The industry will build them because they're simply better cars. The industry won't build them because they don't produce enough profit. Declining oil reserves will make them inevitable. We'll continue to drown in cheap oil. People will love them. People will hate them. There's logic and evidence for all these positions, and they'll all factor in to what eventually happens. As for myself, I'm more sanguine than Ralph Nader or Brad Snell, but not quite as optimistic as Amory Lovins.

CHAPTER 8

JUMP-STARTING THE EV: FEDERAL FUNDING FOR ALTERNATIVE FUEL

UNTIL RECENTLY, THE federal government played a remarkably small role in the auto industry, whose major turning points have almost all been dictated by marketplace forces. It wasn't a government agency that decided gasoline would triumph over steam and electricity at the turn of the century, and no bureaucrat demanded front-wheel drivetrains or disc brakes.

Federal involvement is still largely passive, though the Clinton administration has funded some high-profile (but entirely voluntary) initiatives and student contests that fall far short of the kind of strong mandates some environmentalists would like to see. Having observed how deeply the industry has been affected, even globally, by California's zero-emissions initiative, I was curious to see how state law compared to federal law. I found that, when it comes to Congress, the auto industry is mostly still in the driver's seat.

The most important legislation affecting the auto industry is the federal Clean Air Act, which in 1990 dictated an overall 40 percent reduction in new car hydrocarbon emission levels and a 60 percent drop in nitrogen oxides by 1996. The Environmental Protection Agency dragged out the required update of the emissions standards, prompting a lawsuit from the American Lung Association that prodded the agency to prepare stronger, Tier 2 draft regulations by mid-1997, and finalize them in the spring of 1999. Designed to take effect in the 2004 model year, the standards have won plaudits from some environmental groups, as well as complaints that the heaviest sport-utility vehicles won't be regulated until 2009. "The plan still contains loopholes big enough to drive a Chevy Suburban through," said Ann Mesnikoff, director of the Sierra Club's Clean Car Campaign in February 1999.[1] A 1999 report by the U.S. Public Interest Research Group entitled "Big Cars, Dirty Air" found that if the loopholes Mesnikoff describes were closed immediately, the nation could avoid the creation of 1.2 million tons of smog-related pollution annually.[2] Still, the plan, which also requires sharp reductions in gasoline sulfur levels, was regarded by most environmentalists as the best they could expect from the Clinton administration.

The federal standards in place in 1999 were much weaker than those in effect in California, where low-emission vehicles (LEVs) were allowed to emit only a quarter of the hydrocarbons of a "federal car." California then estimated the cost to automakers of taking a federal vehicle to LEV standards at $120, significant money when spread across millions of cars, though the increased costs are passed on to consumers. The industry's main worry was that an increasing number of states would adopt California's rules and, largely to avert that disaster, it volunteered in 1998 to start producing cars that are somewhat cleaner than the federal standard. It's a clever strategy, because it gives the companies room to maneuver and leaves them in control.

Emissions standards are difficult to enforce under the best of circumstances, because new cars that perform well on paper may not actually do so in the real world. In 1998, Honda and

Ford paid $15 million in federal fines for selling cars with malfunctioning environmental controls.

Though the auto industry hasn't always gotten its way on clean air legislation, its enormous clout has usually allowed it to prevail in Washington. The much-hated CAFE standards, which are federal fuel-economy rules, have not been toughened in over a decade. Even on safety issues, where federal agencies have been more active, the auto industry has largely dictated the timetable.

Clarence Ditlow, head of the consumer advocacy group Center for Auto Safety, calls the federal government "all show and no go" on both safety and emissions. "The government has never been willing to stand up to the car companies, and that's what it takes," he said in an interview. "The agencies have funded prototype vehicles, but that hasn't had much effect on the cars we actually drive. For that we need strong government mandates." Though he didn't mention it by name, Ditlow was casting aspersions on the Clinton administration's automotive showpiece, the Partnership for a New Generation of Vehicles (PNGV).

WORKING TOGETHER: THE PARTNERSHIP FOR A NEW GENERATION OF VEHICLES

With much fanfare, the Clinton administration initiated the PNGV in late 1993 as a government-industry collaboration that would solve the smog problem. President Clinton himself seemed enthusiastic and repeatedly peppered speeches with references to fuel cells and what he twice described as "fuel-injection engines." (Since almost all current cars have such engines, it's safe to assume that the president meant "direct injection," the technology used in the cleaner diesels.) Vice President Gore called PNGV "an historic collaboration."

Working as PNGV, seven government agencies and the United States Council for Automotive Research (USCAR), a consortium with representatives from GM, Ford, and Daimler-Chrysler, had ten years to produce low-emission, eighty-miles-per-gallon family cars. These designs are first appearing as working models in 2000, with a timetable calling for "production-ready" prototypes by 2004, but the program isn't binding. The prototypes may do little more than revolve on show stands.

PNGV is good publicity for both the government and the industry participants. The administration gets a clean-air program aimed at cutting dependence on foreign oil, all on appropriations of about $200 million a year. The USCAR members can derail criticism about their absurdly poor performance on fuel economy. It's an unfortunate fact that the average car going into junkyards today gets better gas mileage than the average car in dealers' showrooms.[3]

Carmakers, in fact, have never even achieved the rather pathetic 27.5 miles per gallon CAFE standard set for their product ranges in 1985 and have paid only very light penalties for producing gas-guzzling sport-utility vehicles, which have been in 30 percent of American garages at one time or another. Influenced by their Detroit competition, import cars have also delivered declining fuel economy. In 1992, the American car buyer could choose from among twenty-six compacts that got at least thirty miles per gallon, but there are only a handful of such cars available today.[4] Domestically, proposals to increase CAFE standards have never gained congressional majorities, though in May of 1999 thirty-one U.S. Senators sent a letter to President Clinton calling for strengthened standards.

Carmakers much prefer voluntary programs to gasoline taxes, though such taxes, now at an all-time low in real dollars, may be the only way to get Americans out of large cars and trucks. According to James Womack, former director of the International Motor Vehicle Program at MIT, gas taxes like those in Europe, which can push a gallon over $4, provide compelling incentives. "Manufacturers are told to make more fuel-efficient cars, but with the real price of gasoline going down, those cars make absolutely no sense to consumers. If this stuff is so cheap, why not use it?"[5]

"In Europe, high gas taxes are the approach of first resort," writes Jane Holtz Kay in *Asphalt Nation.* "In America, however, such taxes have been the argument of first reproach in the political arena. The trust fund built with money from the gas pump remains the piggy bank for the highways."[6] And yet, she notes, California raised gas taxes ten cents a gallon and received few complaints and a lot of benefits.

Carmakers do not voluntarily increase the fuel efficiency of their automobiles. This was explained to me by the very patient Victor Wong of the Sloan Automotive Laboratories at MIT. Although it's housed at the prestigious Cambridge-based university, the Sloan Labs, named for the legendary Alfred P. Sloan, who ruled GM with an iron hand from 1923 to 1937, are actually totally supported by the auto industry. The labs are a research tool for the industry, supplying basic engine design, fueling, and lubrication data.

I asked Wong how clean internal-combustion engines could be, and it turned out to be a naïve question. "I guess the EPA wants to know that, too," he said. "But engines won't be much cleaner than the federal and state emission levels. They are designed when new to produce emissions that are somewhat lower than those standards, so that when they're older they'll meet selective enforcement, in which cars are randomly pulled aside and tested."

Why not, then, make a cleaner car and tout its benefits to the environment? Another naïve question. "You can always make a car cleaner," Wong said. "But if it costs more to make it cleaner, they won't do it. You can be certain that they will not produce cars that produce ten times fewer emissions, because it costs much more and there will be trade-offs somewhere else. Instead, they'll build a car that is 'clean enough.' " And that makes unambiguous regulation all the more important.

But given all this, PNGV's goal of producing eighty-miles-per-gallon cars while meeting emission requirements is certainly laudable. The program also has the worthy aim of reducing the weight of the average family sedan from thirty-three hundred pounds to two thousand pounds, which undoubtedly put wings on the Chrysler ESX2 and Ford P2000 cars, and helped the GM EV1 slim down.

How will Detroit do it? For the initial segment of the PNGV program, direct-injection diesel hybrids have the upper hand. Continuous-combustion engines (gas-turbine and Stirling-cycle power plants) have fallen by the wayside, as have battery EVs and flywheel hybrids. No one is sure how the direct-injection diesel cars will qualify on emissions, since they are major soot emitters, and soot is a principal component in smog. But Ron York, who heads GM's PNGV efforts, defends the diesel, noting that it emits less carbon dioxide than the gas engine, thus helping win the war against global warming. If efforts to remove sulfur from diesel fuel continue, he said, the diesel will get cleaner still.[7]

PNGV is headed by George Joy, not exactly an industry outsider. Joy, who took over as head of the technical task force in 1997, comes to federal service from a long career at Allied Signal, where he worked on automotive gas turbines (ironically, one of the technologies discounted by the PNGV).

Joy told me that a combination of new innovations, from very efficient direct-injection engines to advanced batteries and motors, all placed in very lightweight composite body structures, will help the USCAR designers "get very close to their goal of tripling fuel economy." He acknowledged that it would be difficult to meet that target without using a diesel. "But these are a different class of diesel," he said. "They will have to meet the EPA's Tier 2 standards, which are very tough, allowing about half the emissions permitted by current cars."

The concept cars are just interim designs, of course. The vehicles that will be presented in 2004 will be much more refined preproduction prototypes, engineered to a near-showroom standard, with all the comforts the very demanding American consumer expects. Can such cars be built to an affordable cost, making actual production much more likely? "That's the goal," said Joy. "At year zero, 1993, nobody knew how to get as close technically as we are now. Now the effort is focusing on practicality and cost-effectiveness, and a tremendous amount of work is being done to reach the cost targets."

Fuel cells remain a big question mark. "I'm a little vague on them," Joy admits. "There have been tremendous improvements, and there is certainly a lot of commercial interest. We have tremendous hope and expectations for the fuel cell, but I can't give you a prediction on when they'll be ready. Right now they cost too much."

Will we ever see a time when carmakers are *required* to produce eighty-miles-per-gallon cars? "I doubt it," says Joy. "There's been no change in the CAFE requirement for ten or twelve years. We know how to do it, but it's just very expensive."

PNGV certainly has its critics. The Sierra Club's Dan Becker agrees with Clarence Ditlow that voluntary initiatives are no substitute for government mandates. "We view it as a scam to keep regulation at bay," says Becker, who is otherwise quite supportive of clean-car developments. The PNGV program was no doubt stricken when it was assailed by the prestigious National Research Council, an arm of the National Academy of Sciences. In its third PNGV report, published in 1997, the council announced, "Despite significant progress in a number of critical areas . . . the effort being expended on candidate technologies and systems is not consistent with the likelihood that each will meet performance goals within the program schedule. . . . Work on many critical systems is inadequately funded and lacks integrated technical direction."[8] By the time of the fourth Council report, in 1998, the criticism was more tempered, though it was projected that PNGV would fall short of its eighty-miles-per-gallon goal. The council recommended that more work be done on technologies that might actually provide some public benefit.

Joy defends PNGV as "a way of trying to break out of the way the government and industry worked, which was a command-and-control regulatory structure. Now we're a little more technically driven and not as much regulatory driven." But that's precisely the problem with PNGV, according to consumer advocate Ralph Nader.

"PNGV is a contributor to the auto industry's stagnation, not an answer to it," Nader says. "What Clinton did in 1993 was to set up a partnership that got the auto companies off the antitrust hook so they could collude in the guise of collaboration. It sanctioned them to do what they'd previously been forbidden to do by law. Now they could say, 'There's a breakthrough coming, so why regulate?' It was a very astute maneuver by Detroit."

Nader was talking about a 1969 civil action, initiated by the Justice Department's anti-trust division, against the Big Three automakers for acting together to noncompetitively slow down and suppress the installation of anti-pollution devices in their cars. Nader himself had been a party to bringing the suit, which was settled later that same year by a consent decree.

In 1964, an angry Los Angeles pollution-control officer named Smith Griswold made a speech in Houston, Texas, in which he denounced the auto industry for failing to do its part in the war against smog, then reaching crisis levels in his city. "Everything that the industry is able to do today to control auto exhaust was possible technically ten years ago," Griswold told the Air Pollution Control Association. "No new principles had to be developed; no technological advance was needed; no scientific breakthroughs were required."[9]

Griswold's speech landed on the desk of Ralph Nader, then an obscure Labor Department volunteer. As a trained lawyer, Nader thought he recognized a classic case of "product fixing," and he took the matter to the Johnson administration's Justice Department, then headed by Attorney General Ramsey Clark. Nader's case, reinforced by a condemnatory resolution adopted unanimously by the Los Angeles County Board of Supervisors, was compelling.

A grand jury failed to return a criminal indictment against the automakers but, in 1969, just before Republican appointee John Mitchell took over the Justice Department, a sweeping civil action was filed. The conspirators had tried, the suit alleged, "to eliminate all competition among themselves in the research, development, manufacture and installation of motor vehicle air pollution control equipment."[10]

The automakers, through the trade body that was then called the Automobile Manufacturers Association (itself a defendant), struck back, apparently offended that the department would attack "the industry's 15-year-old cooperative program" that had "succeeded in reducing the level of hydrocarbon emissions of new cars approximately 63 percent below the level of the pre-control models."[11]

The carmakers shouldn't have worried. The Nixon administration let them off the hook with a mild consent decree that allowed them to admit no liability but instead agree not to be conspirators henceforth. "Is this where five years of Antitrust Division involvement and expenditure of numerous man-years is to end?" Nader asked in a 1969 letter to the Justice Department. An important provision of the decree was that the automakers weren't supposed to share one another's proprietary emissions technology, which, of course, is precisely what they're doing as part of PNGV. The consent decree was presumably still in effect when President Clinton joined the auto companies together in one big Three Musketeers' cooperative.

One PNGV insider, Deborah Gordon, doesn't see anything sinister about the companies' collaboration. As director of the Next Generation Transportation Strategies Project at the Yale Center for Environmental Law and Policy, she is an impassioned advocate for public transportation who's trying to reduce the congestion on Connecticut's highways. She's also on the advisory committee to PNGV. "The automakers all come in and present their plans in one big room," she says. "It seems to me that they're still very competitive with each other."

Yes, automakers will always compete for market share, and their interests diverge on many issues, but they've shown a remarkable unanimity on environmental regulation. If the strident positions taken by the carmakers' now-defunct trade association had any internal dissenters, it wasn't immediately apparent. But that, they say, was the dirty past. Now the companies, enthusiastic PNGV partners, are willing to build a new generation of clean cars, but they still want to control the process as much as they possibly can. Even though their rhetoric has softened, they remain a powerful lobby.

THE DEPARTMENT OF ENERGY'S RESEARCH ROLE

PNGV is a many-headed hydra, with the involvement of nineteen federal government labs. One of the most prominent players is the Department of Energy, whose Office of Advanced Automotive Technologies provides two-thirds of the federal support for PNGV research and development efforts. One of the sharpest and best-informed bureaucrats I talked to in the course of writing this book was JoAnn Milliken, the DOE's program manager for fuel-cell automotive programs. Aside from furnishing me with many useful important contacts and background papers, Milliken, a former chemist in the Office of Naval Research, had the broad knowledge necessary to help me understand the government's rather complicated alternative-fuels strategy.

"The DOE's role is to support the research that validates the technology," Milliken said. "It's up to the automakers to use it in vehicles and put it on the road." To that end, DOE has funded research into such pressing challenges as compressed natural-gas storage, direct-injection diesels and new emissions-reducing catalytic converters for them, direct-hydrogen fuel cells (in particular, a 50-kilowatt automotive unit that runs without an air compressor), and the Epyx gasoline reformer. DOE also supports a network of national laboratories, including Los Alamos, in New Mexico, where scientists working on federal grants made important breakthroughs on the PEM fuel cell.

"There had already been years of research on PEM cells as part of the Gemini space program," says Jim Danneskiold, a Los Alamos spokesman, "and in the 1980s we were able to take it to the next level, so it could become commercially viable. We knew that fuel cells would only take off when industry became convinced it could make money on them." Los Alamos worked with General Motors and Ballard on PEM research for eight years, beginning in 1988, resulting in a 10-kilowatt demonstrator that opened a lot of eyes.

I also had an interesting e-mail exchange with Los Alamos scientist Nick Vanderborgh, who wonders why PNGV is focusing on the American family sedan while the actual consumers are lining up to buy sport-utilities. Meanwhile, Toyota puts its fuel cell in the four-wheel-drive RAV4. "This is all very confusing to me," he wrote. "For instance, having a foreign corporation accept and meet the technical challenges assigned to our corporations by our government seems basically strange. It is sort of like somebody in China reading the *Boy Scout Manual,* and doing everything needed to become an Eagle Scout, without being in the scouting program."

DOE's work seems to be designed to give American companies an edge, though the level of support in no way compares to the federal support forthcoming in Germany and especially Japan, where it has become a national priority. In 1993, Japan announced an ambitious, twenty-eight-year, $11 billion hydrogen research program called New Sunshine, which eclipsed Germany's to become the biggest in the world. Basic hydrogen research, including the work on metal-hydride storage that is reflected in Toyota's fuel-cell prototypes, is a major part of the Japanese renewable energy consortium. German government support has declined since reunification, but there was still a $12 million budget in 1995.[12]

The DOE's Milliken is well aware that hydrogen fuel cells have become an international race where all the runners could stumble over a hidden obstacle known as the liquid fuel reformer. The agency's strategy, as reflected in the work it funds at Los Alamos and also at the Argonne National Labs near Chicago, is a fuel-flexible processor, able to reform gasoline, natural gas, methanol, or ethanol at the flick of a switch. "Fuel processing is our number one priority right now. We want to be able to use the existing fuel infrastructure, which will help us get on the road more quickly," Milliken said. She herself pointed out some of the pitfalls of this approach, including an inherently more complicated, heavier system and continued reliance on fossil fuels.

DOE is also funding direct-hydrogen research, including some very promising experiments with fuel tanks pressurized at 5,000 pounds per square inch, which could yield a reasonable range without a reformer. Carbon-fiber composites, also used in lightweight but strong car bodies, have proven to be sturdy enough for such use but are very expensive. The agency's goal is a tank that would yield a range of 340 miles while retaining reasonable trunk space. On paper, a reasonably light fuel-cell car with a 5,000-psi carbon-fiber tank could travel 220 miles before needing to be refueled.

Direct-hydrogen storage is certainly getting close to delivering an acceptable range, but Milliken is still convinced we'll go through a liquid fuel stage. "We already have a gasoline refueling infrastructure that cost hundreds of billions to build," she says. "Why not take a stepwise approach? Go through the reforming stage, then to direct hydrogen."

Though she doesn't think any production-ready fuel-cell car will be built before 2002, Milliken says she finds her work "very exciting. Though I do sometimes ask myself how I'll feel if we fail, and I've spent X number of years on a technology that didn't make it. It's high risk, high payoff."

THE DEFENSE ADVANCED RESEARCH PROJECTS AGENCY

It was in her previous work with the Navy that JoAnn Milliken was first exposed to PEM fuel cells, and that revelation prompted me to ask her why the military is so involved in fuel-cell research. "One of their problems is that they go through a lot of batteries, in the backpacks soldiers use," she said. "They're thinking of replacing those batteries with fuel cells." The military likes fuel-cell vehicles, she adds, because they don't have the "heat signature" that allows internal-combustion engines to be effectively tracked by infrared detectors, and they offer a potentially larger range than present military systems. The Defense Department sees the possibility of ultraquiet scout vehicles and exhaust-free tanks.

Money for alternative-energy projects that might have military applications flows from the Defense Advanced Research Projects Agency (DARPA), which doesn't actually make anything; it just funds research and then gets out of the way. Once the technology is in place for, say, a new generator, it's the navy's job to make sure it's seaworthy and able to stand up to salt spray. DARPA's military mission makes for some strange bedfellows. The agency has poured money into such projects as an electric car designed by a card-carrying Massachusetts environmentalist, a California company's plan to field a utility EV in Hawaii, and a Connecticut-based fuel-cell maker that wants to unhook America from the grid.

There's a clear strategic purpose in all this. DARPA was established forty years ago, after America was surprised by the appearance of the Russian Sputnik satellites. The idea was not to be left behind by technological breakthroughs in the future. (DARPA has sometimes been an innovator itself, as when it created an internal computer network called DARPANET that evolved into the Internet.)

Robert Nowak is a civilian who manages DARPA's advanced energy technology program, and in that capacity he's helicoptered out to aircraft carriers in a quest to find out how American soldiers really use their equipment. "You get an appreciation of what they really need," he says.

Nowak's program supports novel battery concepts, fuel cells, and fuel reforming. "We recognize there's a lot of interest in fuel cells for vehicles," Nowak says, "but we don't want to duplicate existing government efforts. Other agencies are exploring fuel cells for cars. The military has unique challenges, such as the fact that we use diesel and jet fuel. We have to have a reformer for them. The services aren't going to say, 'Let's use methanol' and switch to fuel cells. In the seventies, the army had a successful one-kilowatt fuel cell, but what killed deployment was the logistics of sending another fuel around the world."

DARPA likes PEM fuel cells, and high-temperature solid-oxide cells, too. But diesel fuel is much heavier and more difficult to reform than gasoline, which is difficult enough, so there are major challenges. DARPA funds have gone to Arthur D. Little and Epyx for work on a diesel reformer, and to International Fuel Cells for work with solid oxide, which is very tolerant of high-sulfur military fuels. DARPA also is interested in compact and lightweight mobile fuel-cell power systems for the battlefield, to replace heavy diesel generators. It's a considerable problem to make such systems reliable, since the cells would have to work with very dirty fuels and be "ruggedized" for the field.

Danbury, Connecticut's Energy Research Corporation (ERC), which at one time made silver-zinc batteries for submersible submarines, is now using a DARPA grant to build large-scale molten carbonate fuel cells for distributed power on navy ships. I saw such a unit, complete with diesel reformer, on a visit to Danbury, and it was a rather fearsome, fifteen-foot-tall device wrapped in insulation that operated at 1,200 degrees Fahrenheit. ERC's Peter Voyentzie says, "Putting fuel cells on ships is much easier than putting them in cars." ERC, which makes some of the largest stationary fuel-cell stacks on the market, at two megawatts, is not optimistic about automotive applications. For a new range of EVs to be sold in China and Taiwan, it sells nickel-zinc batteries.

The military is not likely to abandon diesel fuel, but it wants to get away from it in internal-combustion engines. As Nowak points out, the auxiliary diesel power units on military planes produce pollution as they sit idle on airport runways, and in California, the service branches pay a pollution tax on every diesel generator, whether it's used or not.

DARPA is trying to get more user-friendly in the twenty-first century. "By itself, the military is a small customer for alternative energy," says Nowak. "What we want to do is develop technologies that have value to our services, but can be modified for large-scale commercial production, so they benefit everyone. It makes more sense than incurring huge expenses for military-specific technology. This approach works for batteries and fuel cells, not as well for cruise missiles."

COLLEGE ENGINEERS AND THE GOVERNMENT'S FUTURECAR CHALLENGE

For keeping students focused on their work, there's nothing like an electric-car project. The Tour de Sol, an EV race that the Northeast Sustainable Energy Association has run every year since 1989, is frequently won by Massachusetts' Solectria, but there is also keen competition from both high school and college teams. I went to the kickoff in Waterbury, Connecticut, in May 1997 and watched one college team frantically making last-minute adjustments to a hybrid car that had components mounted on every conceivable surface. Big boys Toyota and Ford competed that year, too. The display-only Chrysler Patriot sat forlornly in a corner, and I tried unsuccessfully to find the Toyota representative and persuade him to give me the keys to his very spiffy RAV4 EV. In a seven-day run to Portland, Maine, a good time was had by all, including teams from Boston University, California's Chico State, and the University of New Haven.

Rather less dramatic, but nonetheless an absorbing college project, is the FutureCar Challenge, which is sponsored by DOE and USCAR. The student version of PNGV, it's also a training ground for would-be Big Three engineers, and many contestants get entry-level jobs at Ford, GM, and Daimler-Chrysler after they graduate. "These aren't nineteen-year-olds struggling with freshman English," says spokesman Jack Groh. "They're often graduate engineering students with hands-on support from the auto industry, and they work an unbelievable schedule." In 1997, the Challenge was a race, and then energy secretary Federico Peña waved in the winning team from the University of California at Davis. The 1998 event, which was held at the Henry Ford Museum in Dearborn and involved a lot of performance tests, was fascinating, both for what happened and for what didn't.

Thirteen college teams competed, with many participants hoping that the recognition that would come with winning would buy them a ticket to Detroit. The winners, whose entries were judged on technological innovation, were the University of Wisconsin at Madison, with an all-aluminum hybrid, and Virginia Polytechnic University, with a fuel-cell car.

Well, sort of a fuel-cell car. Virginia Tech didn't actually have a fuel cell, but the spirit was there. "Our plan was to have a fuel cell," says faculty adviser Doug Nelson, a mechanical engineering professor at the school. The idea was to start with a standard 1997 Chevy Lumina bought as a leftover from a dealer's lot, strip it down, and install a 100-kilowatt General Electric motor, a 336-volt Hawker battery pack, and a 20-kilowatt Energy Partners fuel cell from Florida, connected to an onboard carbon-fiber hydrogen tank pressurized at 3,600 pounds per square inch. The car would run on the fuel cell, but the battery would give bursts of power for acceleration, so the car was really a hybrid, with a range of about one hundred miles. At least it was in theory.

The students built the car, trying to save weight as best they could, but there was a slight problem with the fuel cell. It didn't arrive. "It was supposed to come from Energy Partners, but there were technical difficulties with the membrane," says Nelson. "Unfortunately, the cell was a research model, not something we could just go out and buy. And it wasn't ready."

Virginia Tech, a little sheepish, put the fuel-cell car together, with all the plumbing to make the fuel cell work but no fuel cell. Where it was supposed to go they put a box. What they now had was a battery electric car designed to be something else.

Students at Virginia Tech are made of some pretty stern stuff, so they took the car to the competition anyway, and came away with "best acceleration," "best dynamic performance," and "best consumer acceptability." Thanks to some sympathetic judges they tied for first place.

The fuel cell finally arrived months later, in time for the 1999 contest, but there were some teething problems. Without a high-tech reformer, the relatively heavy Lumina had to run on compressed hydrogen. And that meant a range far below the 325 miles specified by the contest. But with a setup as clean as that, who could dispute that the Lumina was a winner on emissions?

What Virginia Tech really wanted was a lightweight aluminum body, like the University of Wisconsin had. This was no ordinary car body: Though it looked just like a production Mercury Sable, it was five hundred pounds lighter. In a program announced in 1990, Ford built one hundred aluminum Sables as test vehicles, some of which it let people drive, others of which it crash-tested. Glenn Bower, a University of Wisconsin mechanical engineering professor who grew up as a farm boy working on tractors and rototillers, managed to talk Ford into giving him one of the leftovers for the Challenge. "I happened to go to a conference, and a Ford PNGV guy was there," Bower says. "I told him I knew about the Sable program, and that I'd heard they were going to crush some of them. Maybe they could make some donations instead." In the end, three Future-Car universities got Sables.

That's not all Bower weaseled out of Ford. He also managed to score an ultrarare prototype compression-ignition, direct-injection diesel engine. Bower definitely wanted a Ford engine under his hood because, the year before, a photo session with Vice President Gore had been abruptly canceled when another school's car was found to have a European engine. This is an all-American contest. "Now I put on my red, white, and blue underwear," says Bower.

Bower's team had fifty to seventy-five students on it, mostly undergraduates, twenty-five of them hard-core workers. To build spirit and greater efficiency, they were divided into groups, with an engine group, a mechanical group, a high-voltage group, and a control group.

There was a lot of cramming, since the team didn't get the Sable until the end of December and the contest was in June. The plan to connect the powertrain wasn't ready until February, so the car had to be built in just four months. The students even built the battery pack, which was made from sixty disassembled 12-volt nickel-cadmium Milwaukee Fat Packs intended for hand-held power tools. The wild concept worked: the handmade battery pack weighed only 125 pounds, when the smallest comparable lead-acid pack was 350.

There were many hassles, but the team got the Sable running, only two days before the contest. The parallel hybrid drivetrain, with a 90-horsepower diesel and a 70-horsepower electric motor working together in apparent harmony, proved to be very efficient and not exactly slow. It outaccelerated a stock three-liter V-6 Taurus, all the while getting great gas mileage. "We were shooting for three times better than the regular car, but we'll settle for two times better," Bower said.

The Wisconsin team came away with the "innovations in

aluminum" award, presented by, what else, the Aluminum Association. It's hardly surprising because, in addition to the aluminum body, the students saved even more weight by hand-fabricating many auxiliary parts in aluminum, from the engine mounts to the suspension uprights. Even the four-piston racing calipers for the brakes, which saved eighteen pounds per corner, were made of composites.

In the hurly-burly of the event, going around the Chrysler test track, the car broke down continuously, and everyone worked until midnight to fix it, even pulling out the whole drivetrain. When a half-shaft broke, they raced down to a Ford dealer and got a stock unit in place just before the qualifying round. There were times when just about everyone wanted to quit, but they didn't. The University of Wisconsin at Madison tied for first place.

In 1999, the team brought the same car back, with a new and even more efficient diesel engine (from a European Ford Focus) and 150 pounds in weight savings. Just by taking some plastic moldings off the doors, they stripped ten pounds per side. The result was an even more competitive car that will get more than seventy miles per gallon on the highway and cruise for eight hundred miles before needing a fill-up.

The hard work and all-nighters paid off. The University of Wisconsin team won the 1999 FutureCar Challenge by a wide margin, and Virginia Tech came in second, this time with its fuel cell in place. "We needed more time and didn't have everything working perfectly," admits Nelson, "but with a fuel cell running on hydrogen gas, we won the emissions category outright." In 2000, FutureCar was replaced by FutureTruck, and both schools began modifying big Chevy Suburbans, aiming to create that genuine novelty, the fuel-efficient SUV.

Neel Vasavada, a Wisconsin mechanical engineering and philosophy major, was set to finish school in 2000 if he didn't take time off again, as he did one semester for a job as a motorcycle mechanic. Vasavada cheerfully admits to being a gearhead; he started entering cars in autocross events when he was sixteen.

He didn't know or care anything about EVs, but after he arrived at Wisconsin and started working in the auto shop, he noticed an EV in the next bay. "I was a racer, and this thing wasn't very interesting," Vasavada said. "It was slow and heavy. But what they were doing with the technology was fascinating, and I surprised myself by starting to work on EVs." And he joined the FutureCar team.

"We built the Sable as a sports sedan," Vasavada said. "Nobody will buy EVs if they're not better than what we have on the market now—not just as good, better." Vasavada took the program's Ford Escort wagon on an autocross. It had only 90 horsepower with an engine built for an irrigation pump, but a lot of the low-end torque that is inherent in EVs. "It was flying," Vasavada said, "and we kicked butt."

Now Vasavada is a convert. He's convinced that hybrid cars can give us not only high efficiency and low emissions, "but also a superior automobile in every sense." The Sable uses the electric drive to get off the line, then the 90-horsepower diesel kicks in for cruising, delivering a total 175 horsepower when they're both on line. Vasavada lays rubber in first gear. He tells people the car is an EV, then watches their jaws drop as he blows them away. He gets fifty-five miles per gallon and still has seating for five, but he also wants to take on a high-performance SHO Taurus—and shut it down. "They shouldn't just market EVs to gray-haired environmentalists," said Vasavada. "They can also appeal to young sports-car fans like myself. EVs can be anything you want them to be, and they don't have to stodgy and slow." Maybe Vasavada will take his EVs racing. He could have shown Chrysler's Patriot team a thing or two. As far as his career goes, he'll probably end up talking to the Big Three recruiters who hang around. "They definitely schmooze us," he said. But before he could put on his suit and meet the car guys, he had to put the Sable back together.

Without question, the federal government could be doing more to encourage alternative-fueled transportation. Political considerations have stymied improvements in national fuel-efficiency standards, as well as any serious proposals for a substantial gasoline tax. The Clinton administration has pushed for neither of these things, which could help end America's fascination with large sport-utility vehicles. Instead, it has showcased its PNGV program and modest tax deductions for clean cars. The EPA does deserve credit, however, for sticking to reasonably strict Clean Air Act regulations, in the face of the usual industry lobbying. The threat of global warming looms over any future policy decisions, but the Potomac may have to overflow its banks before Washington's bitter partisanship on the issue is penetrated.

CLEARING THE AIR: CLEAN CARS AND SUSTAINABLE TRANSPORTATION IN THE TWENTY-FIRST CENTURY

NEAR THE END of my long road trip investigating clean-car progress, I sat in the Connecticut office of Bill Hahn, the big, ruddy Irishman who is vice president of International Fuel Cells (IFC), a division of United Technologies, one of the world's biggest defense contractors. Hahn had just told me he expects that, by 2010, 10 percent of all the cars on the road will use some form of alternative energy.

I'd gotten used to hearing this kind of thing from green-power partisans, but IFC is nothing if not a mainstream American company. As we spoke, the nearby shop floor rang with the sound of workers preparing what was to be a large manufacturing facility to make PEM fuel cells. With Japanese partner Toshiba, IFC (already a leader in building large stationary fuel-cell power plants) was getting into the auto business in a dramatic way.

IFC is no fuel-cell newcomer. In an anteroom, the company proudly displays models of the fuel cells it built for a succession of space explorations, beginning with the Apollo missions in the 1960s. Fuel cells solve two problems for America's space travelers: They offer a completely dependable source of electricity, and they produce drinking water for the crew that would otherwise have to be stored on board. But the Apollo cells are museum pieces now, and even the more modern units carried on five space shuttles are old technology. Today, an Apollo-sized cell could produce 100 kilowatts, fifty times what its predecessor could have.

Visiting IFC was eye-opening, because it gave me a chance to see how serious American industry is becoming about fuel cells and about alternative energy in general. Such a commitment is absolutely necessary, because clean cars can't be legislated into existence. Like Plug Power, Ballard, Energy Partners, and many other companies, IFC wants to be part of what is shaping up as a major commercial enterprise.

The full contours of that enterprise are still not clear, nor is the time line. In the summer of 1998, while I was on the road, a long-awaited report prepared for the California Air Resources Board called *Status and Prospects of Fuel Cells as Automotive Engines* was released. While the report offers a rosy outlook for methanol fuel-cell stacks in cars, it's less sanguine that a direct-hydrogen infrastructure will emerge soon. "Hydrogen is not a feasible fuel for private automobiles now or in the foreseeable future because of the difficulties and costs of storing hydrogen on board and the very large investments that would be required to make hydrogen generally available," the report's authors, members of the board's fuel-cell advisory panel, conclude.[1]

The authors certainly think the automotive fuel cell is coming, for much the same reasons outlined in this book. Noting the $1.5 to $2 billion international investment, they write, "The most compelling argument for the Panel's cautious optimism about the prospects for ultimate success is that the promise of fuel cells as an environmentally superior and more efficient automobile engine is being pursued with an unprecedented combination of resources by powerful organizations acting in their own interest and with strong public support."[2]

The researchers are pessimistic that hydrogen, even compressed at 5,000 pounds per square inch, can be sufficiently downsized into an automobile. Citing a Ford study, they say that even with a fuel efficiency of seventy miles per gallon, the tank necessary to give the car a 350-mile range would seriously compromise both passenger and cargo space. The only alternative is to put the tank on the roof, like the NECAR II van, but no one thinks this is an acceptable alternative in a passenger car. Storing the hydrogen in metal hydride offers some hope, and is being pursued by Toyota and Honda, but the metals are heavy and the costs high.

Two researchers at Northeastern University, Nelly Rodriguez and Terry Baker, say they have developed a system based on the absorption powers of carbon nanofilters for high-density storage of hydrogen. If it worked, it would probably make direct-hydrogen cars practical. And researchers at National University in Singapore are reporting promising results, according to *Science* magazine.[3]

The second obstacle is generating hydrogen and delivering it to customers. The Argonne National Laboratory says that building a production capacity for the United States would be extremely expensive, $10 billion by 2015 and $230 to $400 billion by 2030. Building distribution would add $175 billion by 2030.

Sandy Thomas of Directed Technologies, a Ford consultant, thinks that hydrogen could be delivered at around the same cost as its equivalent in gasoline, but his figures compare a 24.5-miles-per-gallon gasoline car running on taxed gas with an eighty-miles-per-gallon fuel-cell car running on untaxed hydrogen. If both vehicles got eighty miles to the gallon and neither fuel was taxed, hydrogen would cost two to three times more per mile.

What about generating hydrogen through renewable sources? The CARB report seems to assume that hydrogen would be produced at large, centralized facilities, the way gasoline is refined now. But what if hydrogen is made locally at the neighborhood gas station, or in renewable energy "farms"? Princeton professor Joan Ogden has studied the Los Angeles area, and she thinks there is excellent opportunity for solar photovoltaics in the desert areas east of the city. Enough hydrogen could be produced through solar power in an area of twenty-one square miles to fuel a fleet of a million fuel-cell cars on an ongoing basis. And wind sites at Tehachapi Pass and San Gorgonio could achieve similar results, she says.[4] Geothermal power is available, too. The major drawback is that generating hydrogen this way requires long-distance pipelines, and the gas is notoriously leaky.

While renewable-hydrogen demonstration projects are coming, the first commercial fuel-cell cars will almost certainly run on liquid fuels. It's hard to put a date on the emergence of commercial fuel cells and not look like a fool, since the timetables announced by government and industry have generally been proven too conservative. But many auto companies already have running, driving fuel-cell prototypes, and it appears likely that some modest commercialization will be achieved by 2004. Hybrids, well, they're already here.

THE METHANOL FUEL-CELL SCENARIO

From an environmental point of view, how polluting is a fuel-cell car running on methanol? Daimler-Benz was the first company to compile working data, produced from studying

NECAR III emissions on a dynamometer programmed for a typical mix of urban and suburban driving. Initial results were encouraging. There were zero emissions for nitrogen oxide and carbon monoxide, and extremely low hydrocarbon emissions of only .0005 per gram per mile. Worryingly, NECAR III did produce significant quantities of the global-warming gas carbon dioxide, roughly equivalent to the emissions of a direct-injection diesel.

Building a methanol infrastructure may not be all that difficult. The fuel is most easily produced from natural gas, but it can also be distilled from coal or, raising the hopes of renewable advocates, biomass. In the 1980s, methanol had a brief vogue as an internal-combustion fuel, and President George Bush extolled its virtues in a 1989 speech. But methanol is highly toxic, and burning it, while producing slight emissions benefits, also adds substantial amounts of formaldehyde to the witches' brew coming out of the exhaust pipe.

Although the current world methanol network is the equivalent of only 6 percent of U.S. gasoline consumption, new resources could be built quickly. Major manufacturer Methanex told the CARB panel that it could build a $350 million plant in three years that would meet the needs of 500,000 fuel-cell cars. And methanol can be pumped from existing gas stations, though the fluid is corrosive, and pumps, lines, and tanks would have to be remade in stainless steel. But if there's a demand, those costs would likely be borne by the private sector.

Gasoline reforming is still in an embryonic stage, and its inherent complexity has discouraged Chrysler and other companies that once championed it. A major problem is the presence of sulfur in the pump grades. Sulfur is poisonous to the catalysts in the PEM fuel cell. One answer to this is a zinc-oxide bed to capture the sulfur, but that's likely to add to the expense, weight, and size of the reformer. More likely, refineries will be asked to produce a new grade of gasoline with near-zero sulfur content. That's a significant public benefit, since sulfur in gasoline today is a health threat.

In addition to being a smog enhancer, sulfur inhibits the performance of internal-combustion catalytic converters in the same way it affects fuel cells and it can increase emissions by as much as 20 percent. California legislates low-sulfur fuel, and the average there is 40 parts per million, while in the rest of the country it is about 350 parts per million. The oil companies are resisting a national low-sulfur standard, saying it would cost too much, but CARB estimates that cost at only five cents per gallon.[5]

The car companies and their Big Oil counterparts would love to see fuel-cell cars running on gasoline, since that scenario represents minimum disruption for them, but it isn't likely to happen. And the prediction that methanol will serve as a bridge to direct hydrogen—the automakers' second-favorite scenario—could be wrong, too. Though early fuel-cell cars will mostly run on methanol, unexpectedly rapid advances in direct-hydrogen storage and production may make any liquid fuel redundant.

REAPING THE BENEFITS OF THE NEW AUTOMOTIVE RESEARCH

Fuel-cell cars won't become kings of the road overnight, and nobody will be able to accurately date "the death of internal combustion." In the meantime, the benefits of alternative-fuel research are already being felt. Consider the steel industry. Here's a bedrock American institution whose opinion was not consulted when automakers started talking about all-composite and all-aluminum bodies for their lightweight cars. The use of plastics in cars has become increasingly widespread since Henry Ford tested his soy panels and Chevrolet introduced the fiberglass Corvette in 1953. Most EV prototypes are made of composites, and the production EV1 uses them in a mix with aluminum. But faced with this challenge, the steel companies didn't sit idly by. Using light but high-strength steel and fewer parts, they produced a prototype ultralight steel auto body that could lower the weight of a finished car by 25 percent.

"The greatest thing that ever happened to steel was plastics," said Darryl Martin, automotive director of the American Iron and Steel Institute. "It kicked us in the ass."[6] In 1997, the steel industry launched a five-year, $100 million ad blitz to bolster its image.

According to Joe Carpenter, a DOE employee who is PNGV's point man on lightweight materials, the average family sedan today weighs thirty-three hundred pounds (reflecting an actual weight gain since 1982), but the PNGV goal is two thousand, meaning significant dieting is necessary. Car bodies could get both lighter and stronger, which is one of the attractions of composites. "I think there will be a horse race over the next twenty-five to thirty years between the various options," Carpenter said, "and that includes steel, which is by no means out of the picture. The steel people are worried, though, and that's why they got together this $24 million ultralight project."

THE RESPONSIBLE SUV?

Making cars lighter is a significant social benefit, and so is making them more fuel efficient. If, as many carmakers hint, the American sport-utility vehicle will be the first in line to get the hybrid treatment, we'll all be better off, because today's sport-utilities are very bad indeed, averaging just seventeen miles per gallon. And they're also unsafe—at least for the other drivers on the road. When large sport-utilities hit cars head-on, according to a 1998 University of Michigan study, the occupants are five times more likely to die than if a car had hit them.[7]

Since 1990, reports Friends of the Earth on its "Roadhogs" web page, Americans have wasted an extra seventy billion gallons of gasoline because of sport-utilities. The off-road behemoths have been permitted by federal law to waste 33 percent more gasoline than passenger cars, and also spew out 30 percent more carbon monoxide and hydrocarbons and 75 percent more nitrogen oxides.[8] A 1999 survey of SUV owners by the U.S. Public Interest Research Group found that only 17 percent of respondents were aware that their vehicles polluted more than cars.

In 1997, I was along for the ride when the Dodge Durango met the press in a tent outside the Four Seasons Hotel in Austin, Texas. Auto journalists fresh from enjoying the free buffet made the appropriate noises as this monument to conspicuous consumption—a 245-horsepower, V-8-powered behemoth that seats eight and gets twelve miles per gallon—prepared to join a field already crowded with thirty-seven other largely impractical sport-utility vehicles. The same Americans who twenty years earlier worried about America's dependence on foreign oil bought 2.4 million sport-utilities in 1997, up from 1.1 million in 1992.

With Dodge making a hefty $8,000 profit on every Durango, unrelenting consumer demand, and the ready availability of gasoline at a third the price Europeans pay, there's been little reason until now for the car companies—or consumers, either—to be thinking about energy conservation or downsizing the product line. From the vantage point of that tent, the raging debate about the ethical car seemed almost surreal.

It's amusing to note the existence of an industry-sponsored group called Tread Lightly!, which urges sport-utility owners to be careful when driving their vehicles off public roads. It makes good public relations (and a vivid contrast to the industry's own ads showing the vehicles in the middle of creek beds) but the advice is quite unnecessary. A CarPoint survey reports that fewer than 10 percent of all sport-utility vehicles ever leave the road, mainly because their owners are afraid of scratching them.[9] The protective attitude tips off a passion for these vehicles that is quite astonishing. Conservative syndicated columnist Cal Thomas, afraid that safety concerns will take the sport-utility right out of his family driveway, rails, "First our guns, now our sport-utility vehicles! . . . They'll get my wife's sport-utility when they pry her fingers off the four-wheel-drive shift!"[10]

But if Americans aren't going to willingly climb out of their sport-utilities, perhaps the sport-utility can be improved. That's already beginning to happen. The Mercedes ML320, for instance, was designed with bumpers at the same height as a standard passenger car's, so it's less likely to cause fatal damage in a head-on crash. Prodded by William Clay Ford, all Ford sport-utilities built after the 1999 model year meet California low-emission vehicle standards.

A hybrid sport-utility of the type some manufacturers are talking about would be a big improvement. Carmakers need look no further than the hybrid electric, camouflage-painted Hummer that's been turning up at EV shows. Built by PEI Electronics in Huntsville, Alabama, the DARPA-funded Hummer outdragged a stock model, yet gets much better gas mileage with lower emissions. In pure electric mode, this military vehicle can't be located by an enemy's heat-detecting equipment.

DO CLEAN CARS MISS THE POINT?

Sport-utility vehicles can be downsized and their emissions controlled. Hybrid cars can rack up impressive fuel economy. And fuel-cell cars can eliminate the automotive tailpipe altogether. But is that enough? Isn't it cars themselves that are the problem? For one hundred years, we've defined the automobile as thirty-five hundred pounds of steel rolling down the road on rubber tires. In that century we've devised huge improvements in technology to make them go faster, ride smoother, and be far more comfortable and also much safer than they've ever been before. But the basic equation hasn't changed. Cars, with their sheer numbers increasing geometrically, demand more and more of our physical space, for highways, parking lots, and garages.

It's not hard to imagine, in 2025, that we'll have legislated the polluting tailpipe out of existence but done precious little to end the gridlock that, in some major traffic corridors, threatens to make "rush hour" a twenty-four-hour phenomenon. In the not-too-distant future, will we have replaced the idling rows of fossil-fuel burners with a new plastic bumper–to–plastic bumper lockup of "clean cars"?

It's quite likely that we will. Despite growing awareness of problems such as global warming and a willingness, expressed in many opinion polls, for a fuller range of mass-transit options, Americans are depending on their gas guzzlers more than ever. Since 1983, average per capita fuel use for transportation has risen 12 percent, returning us to pre-"energy crisis" levels. One in five households today has three or more cars (it was one in twenty-five in 1969). The industry equation that more muscle equals stronger sales has meant that these cars get more powerful every year: Average horsepower grew almost a third (from 99 to 156) in the fourteen years between 1982 and

1996. Suburban sprawl also propelled a gradual increase in commuting distances, which grew by a third (to an average of 11.6 miles) from 1983 to 1995.[11]

Is there a twelve-step program for three-car-garage families? Can we cure our auto addiction? It may be a profoundly desirable goal, but it won't be easy. Despite many visionary projects and an evolving consciousness about how to make mass transit work—by building interconnected bus and train hubs, for instance—these services are facing *declining* patronage from America's auto dependents.

During World War II, when fuel was rationed, trolleys and buses carried 25 billion passengers a year; by the early 1990s, that figure, despite a huge population increase, had dropped to nine billion trips.[12] Today, just 2.5 percent of American trips are taken in all forms of public transportation, including buses, subways, light rail, trains, planes, and even taxis. Americans use twice the energy per person of Europeans, who drive half as much and use mass transit as much as ten times more.[13]

Given those depressing statistics, car-phobic environmentalists may need to recognize the major gains possible in simply cleaning up the tailpipe. Getting cars off the road is a Herculean effort, made even harder by population increases. In Fairfield County, Connecticut, for instance, where an interstate built for 50,000 vehicles a day actually carries 150,000 and traffic builds at 1.5 percent a year, beleaguered state planners, working with industry and academia in a consortium known as the Coastal Corridor Coalition, say the best they can hope for is a five-year, 5 percent traffic reduction. The state will try to get commuters out of their cars with a series of modest incentives designed to bolster rail and bus travel, as well as encourage van pooling, flexible work hours, and telecommuting. Even if it works, the plan only forestalls complete gridlock: The highways will be almost as congested as before. And Fairfield County *has* mass transit. While intercity bus service is poor or nonexistent, most commuter communities are served by Metro North trains.

There is considerable merit to the idea of using the urban planning process to lessen auto dependence, and Vancouver, British Columbia, is an excellent example. Through a combination of moves to encourage pedestrians and "calm" traffic (wide sidewalks, pocket parks, bike racks, road diverters, mixed-use high-rise residential and office towers, and crime reduction), pedestrians have "taken back" the West End of the city. City Council member Gordon Price counts ten pedestrians for every moving car and isn't afraid to stand in the center of the street on a busy Saturday. "We're in the middle of the highest-density residential area in western Canada, and we're not even thinking about traffic," he says.[14]

Asphalt Nation author Jane Holtz Kay says "there's no such thing as a clean car," and given their overall impact on the planet, she's got a point. To some environmental purists, tinkering with automotive technology is like rearranging the *Titanic*'s deck chairs. Alan Thein Durning of Northwest Environment Watch calls the automobile "the proximate cause of more environmental harm than any other artifact of everyday life on the continent."[15] In his book *End of the Road,* Wolfgang Zuckerman argues that the clean car "would do very little to solve the problems of congestion, traffic casualties, aesthetic pollution, and destruction of cities and countryside. . . . We have to make do with fewer cars."[16]

Unfortunately, the plight of those planners in Fairfield County makes plain how difficult it is to reduce real-world auto travel. As Lewis Carroll put it, sometimes it takes all the running you can do to stay in the same place. America's automotive "population" grows 5.3 percent a year. Against that rising tide are some modest but innovative programs. Car sharing, either as part of an informal arrangement or as a formal business proposition, has proved quite successful, especially in Europe. Under the now-defunct Short-Term Auto Rental (STAR) program in San Francisco, residents of a large apartment and condominium complex had twenty-four-hour access to a fleet of communal cars, paying a monthly fee based on usage. In Berlin, Germany, the Stattauto car co-op runs a network of fourteen lots around the city. Borrowers make reservations, then pick up their keys from safety-deposit boxes, ultimately returning the car to one of the fourteen lots. Honda has developed a similar program called the Intelligent Community Vehicle System (ICVS). Users of the system, which has yet to be implemented, are issued cards that plug into special computer ports at the ICVS stations; their bank accounts are automatically debited for the time they have the car. ICVS vehicles are tiny one- and two-passenger battery and hybrid-electric city cars, plus electric bikes. I was able to see the system in action on a 1999 visit to a Honda facility in Motegi, Japan, and it's wonderfully efficient.

Daniel Sperling, director of the Institute of Transportation Studies at the University of California, Davis, is helping launch a car-sharing program in northern California. Drivers will be able to drop off a co-op car at rapid-transit stops, or at major job destinations like Lawrence Livermore Labs. "The evidence from Europe shows that people who take part in car-sharing programs end up using much more public transportation," Sperling says.

Despite his clear role as part of the solution, however, Sperling is pessimistic that American auto dependence is going to end anytime soon. "Given our present technology, that's something of a lost cause," he says. "Unless there's a dramatic redesign of transit services—like very fast and efficient trains—or a sudden turnaround in land use, cars are the name of the game. It's foolhardy to focus all our environmental efforts on the two or three percent of the population riding mass transit and ignore the more than ninety percent driving cars."

That's also the majority view at the Sierra Club. Energy campaigner Dan Becker says, "We need to change America's love affair with the car, but first we need to change the car. It's hard to influence the thinking of 240 million Americans, but there are only twelve companies that make the bulk of the world's cars. It's much easier to turn them around." Becker sees "dramatic changes for the better" on the horizon and has ample praise for the hydrogen fuel cell. "It emits water," he says. "We like water."

Even an incremental improvement in how cars perform makes a huge difference. According to Sierra Club figures, a car getting only eighteen miles per gallon (typical of sport-utility vehicles today) will produce fifty-eight tons of carbon dioxide in its on-road life, while a car achieving forty-five miles per gallon will produce only twenty-three tons.[17] The auto industry knows how to produce cars that get forty-five miles per gallon, and it will do so if there's sufficient public demand. It's certainly a useful interim goal, if not a long-term solution to our serious transportation problems.

READY OR NOT . . .

A number of things could happen that would derail the clean car. Automotive tailpipes could be exonerated as a factor in global warming. Smog could lift in many major cities, putting the brakes on clean-air laws. Public transit could blossom, making the private automobile less of a factor in the transportation grid. The sport-utility fad could die. Technical difficulties could prove insurmountable to the commercial development of the fuel cell. Huge new supplies of cheap oil could be found. Two-dollar-a-gallon gas taxes could lure Americans into tiny Euro-sized cars with low-emission engines. The California Air Resources Board, the single biggest legislative driver, could cave in on its mandate requiring that 10 percent of an automaker's fleet be zero-emission vehicles by 2003, as it did on the 2 percent by 1998 regulation. Congress could repeal the Clean Air Act.

I don't think any of these things will happen. America won't change overnight, and we will need clean cars. I've traveled across the country, to Europe, and to Canada, and I've seen some truly remarkable competition-driven achievements. The fuel cell, with a $2 billion international research commitment, has come very far, very fast, and promises to revolutionize not only the personal car but the entire energy economy.

Although the auto industry has shown an incredibly consistent dedication to maximizing its shareholder profit, to the extent that it declines to spend an extra nickel per car on safety latches it also recognizes what looks to be a quantum leap forward. As Chris Borroni-Bird told me, the fuel-cell car, by delivering range, near-zero emissions, and unprecedented efficiency, can potentially meet the industry's three most important goals.

Carmakers aren't eternally married to internal combustion, though they've had a glorious ride for the better part of one hundred years. The technology, with its lucrative parts and service businesses, has served the auto industry well, but the speed with which it can be abandoned, given something better, may astound some observers. Within five years of the first motor vehicles, the horse was ready for an honorable retirement.

A similar transition may be ahead, though it will probably take longer than five years. Speaking at a forum on fuel cells at the 1999 New York International Auto Show, Neil Ressler, Ford's chief technical officer, was unequivocal about the need for clean cars, even in the absence of a gas crisis or overwhelming consumer demand. "Ford believes that EVs have to happen, because the environmental drivers are there," he said. "If we have to be out ahead of the public on these cars, then we will be."

Readers looking for heroes probably won't find them in the car industry. The leadership, as Michael Moore so memorably demonstrated in the film *Roger and Me,* is guided by bottom-line thinking and has historically lacked vision. To make change, it needs a good push. But there are good, sound business reasons to justify EV investment, as well as an international race that could be devastating to any company not already at the starting line.

We're entering an exciting period of transition. As I was finishing this book, I finally persuaded Toyota to lend me a Prius hybrid for a couple of days, without anxious handlers or route limits. Here was my chance to test the car of the future under real-world conditions. I expected that using it for picking up groceries and the kids from school would require some period of adjustment. After all, the Prius makes a revolutionary statement, so how could I expect it to be easily domesticated? What if I hated it?

But the Toyota made it easy for me, especially after I got used to the right-hand drive of the Japanese-spec car (the American model, of course, would have left-hand drive). Despite incorporating nearly all of its two drivetrains under the hood, the Prius handled well and never felt front-heavy. The back seat offered plenty of legroom for my wriggling kids, and the trunk was capacious. Power windows and locks, plus frigid air-conditioning, meant that I was hardly roughing it. And two days of near-constant driving hardly moved the fuel gauge. When the car's computer switched the gasoline motor on or off (as it did frequently), there was a tiny shudder. That little jolt is the small sacrifice Prius owners will need to make if they're going to drive a cleaner car.

With its driver on the right and its unfamiliar shape, the Prius turned heads. "Is that the half-gas, half-electric car?" a passerby asked me. "Yes it is," I answered, but it's also something more: the lead vehicle in what is soon to be a parade of environmentally oriented automobiles. I hope we're ready for them.

PEOPLE INTERVIEWED

Steve Brill, editor and publisher of the media review *Brill's Content,* recently complained that authors boast of interviewing "hundreds" of people without specifically enumerating them. For the record, then, I did ninety-eight full-length interviews, with the following people:

Sandors Abens, director of engineering and transportation systems, Energy Research Corporation
Mark Amstock, national alternative fuel vehicle planning manager, Toyota
Chris Ball, outreach director, Ozone Action
Dan Becker, director, global warming and energy programs, the Sierra Club
Fred Beer, EV1 driver
Ed Begley Jr., actor and EV1 driver
Robert Beinenfeld, senior manager, Honda Alternative Fuel Vehicles
Olle Boethius, alternative fuels program, Volvo
Chris Borroni-Bird, manager of technology strategy, Chrysler
Glenn Bower, professor of mechanical engineering, University of Wisconsin, Madison
James S. Cannon, president, Energy Futures, Inc.
Joe Carpenter, Partnership for a New Generation of Vehicles lightweight materials manager, Department of Energy
Mike Clement, manager, alternative vehicle sales and marketing, Chrysler
Csaba Csere, editor in chief, *Car and Driver*
Jim Danneskiold, Los Alamos National Laboratories
David E. Davis, editor, *Automobile*

Bob Derby, spokesman, Epyx Corporation
Clarence Ditlow, executive director, Center for Auto Safety
Tom Dowling, EV1 driver
Johannes W. Ebner, vice president, Fuel Cell Project House, DaimlerChrysler
Gerald A. Esper, American Automobile Manufacturers Association
Nancy Floyd, co-founder, Nth Power Technologies
Art Garner, manager, Honda public relations
Robert Gerber, Ford Ranger EV owner
Deborah Gordon, director, Next Generation Transportation Strategies Project, Yale Center for Environmental Law and Policy
Charles Griffith, auto project director, Ecology Center of Michigan
Jack Groh, spokesman, FutureCar Challenge
William C. Hahn, vice president, International Fuel Cells
Tim Hastrup, Honda EV Plus driver
Nancy Hazard, director, Northeast Sustainable Energy Association
Fred Heiler, director, Mercedes-Benz public relations
Hazel Henderson, futurist
Dave Hermance, general manager, Powertrain, Toyota Technical Center
Peter Hoffman, editor, *Hydrogen and Fuel Cell Letter*
Gerald Hornburg, manager, fuel-cell systems, DBB
Chris Howard, general sales manager, Paragon House of Honda
Roland Hwang, Union of Concerned Scientists
George Joy, chairman, Partnership for a New Generation of Vehicles technical task force
Annette Kliem, new technology press, DaimlerChrysler
Jamie Knapp, California Electric Transportation Coalition

Dr. Richard Krauss, Fuel Cell Project House, Daimler-Chrysler
Christina Lampe-Onnerud, director of energy storage, Bell-core
Paul Lancaster, vice president, corporate development, Ballard Power Systems
Pete Lord, Honda EV Plus driver
Amory B. Lovins, co-CEO (research), Rocky Mountain Institute
Iain MacGill, senior policy analyst, Greenpeace
Richard Maddaloni, project manager, Plug Power
Anita Mangels, executive director, Californians Against Hidden Taxes
Cecile Martin, California Electric Transportation Coalition
Darryl Martin, automotive director, American Iron and Steel Institute
Jerry Martin, California Air Resources Board
Robert Massie, executive director of the Coalition for Environmentally Responsible Economics
J. Byron McCormick, co-director, General Motors Global Alternative Propulsion Center
Dan McCosh, automotive editor, *Popular Science*
Bruce Meland, editor, *Electrifying Times*
Alfred Meyer, manager, transportation business unit, International Fuel Cells
JoAnn Milliken, program manager of fuel-cell automotive programs, Department of Energy
Gary Mittleman, president and CEO, Plug Power
Ralph Nader, consumer advocate
Douglas Nelson, professor of mechanical engineering, Virginia Polytechnic University
Jay Neutzler, program manager and senior research engineer, Energy Partners

Robert Nowak, program manager, Defense Advanced Research Products Agency

Joan Ogden, Center for Energy and Environmental Studies, Princeton University

Hans-Olov Olsson, president and CEO, Volvo Cars of North America

Stanford Ovshinsky, president and CEO, Energy Conversion Devices

Ferdinand Panik, senior vice president, Fuel Cell Project House, DaimlerChrysler

Frank Pereira, advanced technology brand manager, General Motors

John Petersen, director, Arlington Institute

Don Prohaska, Analytic Power Corporation

Robert Purcell Jr., executive director, General Motors Advanced Technology Vehicles

Firoz Rasul, president, Ballard Power Systems

Ron Reid, research engineer, Schatz Energy Research Center, Humboldt State University

Dr. Michael Reindl, chemist, fuel-cell development, Ballard Power Systems

Debby Roman, Ballard Power Systems

Susan Romeo, director of marketing, CALSTART

Susan Rosenberg, United Parcel Service

Marvin Rush, EV1 driver

Melanie Savage, CALSTART

Virginia Scharff, author, *Taking the Wheel: Women and the Coming of the Motor Age*

Peter Schwartz, chairman, Global Business Network

Robert W. Shaw Jr., investor, Arete Corporation

Brad Snell, author

Daniel Sperling, director, Institute of Transportation Studies, University of California at Davis

Robert Stempel, former chairman, General Motors, and senior consultant, Energy Conversion Devices
Karl Thidemann, Solectria
Sandy Thomas, vice president and director of research, Directed Technologies
Dick Thompson, manager of advanced technologies, GM Advanced Technology Vehicles
Nancy Todd, vice president, Ocean Arks International
Kris Trexler, EV1 driver
Jeanne Trombly, manager, West Coast operations, Investors' Circle
Joanna Underwood, president, INFORM
Neel Vasavada, student, University of Wisconsin, Madison
Peter Voyentzie, director of corporate services, Energy Research Corporation
John Wallace, director, advanced fuel vehicles, Ford Motor Company
Brett Williams, senior research associate, Rocky Mountain Institute
Ron Williams, former publisher, Detroit *Metro Times*
Dr. Victor Wong, Sloan Automotive Laboratories, MIT
James Worden, president, Solectria

NOTES

Chapter 1 PULLING THE PLUG: A BRIEF HISTORY OF ALTERNATIVE MOTION

1. Richard Crabb, *Birth of a Giant: The Men and Incidents That Gave America the Motorcar* (Philadelphia: Chilton Book Co., 1969), pp. 17–18.
2. Enzo Angelucci and Alberto Bellucci, *The Automobile from Steam to Gasoline* (New York: McGraw-Hill, 1974), pp. 13–14. The authors also report that a spring-driven car was produced in Nuremberg, Germany, the cradle of watchmaking, circa 1649. It needed winding every five minutes and could travel all of 750 feet.
3. Gary Levine, *The Car Solution: The Steam Engine Comes of Age* (New York: Horizon Press, 1974), pp. 15–16.
4. Lord Montagu of Beaulieu and Anthony Bird, *Steam Cars: 1770 to 1970* (New York: St. Martin's Press, 1971), p. 132. The authors are skeptical that Watt's car would ever have worked. "It is easy to see that the cumbersome machinery of James Watt's steam carriage would have been so heavy, despite the wooden boiler, and have taken up so much space that, had the vehicle moved at all (which is by no means certain), it would have been hopelessly uneconomic" (p. 16).
5. Levine, *The Car Solution,* pp. 17–18. Richard Trevithick built two steam carriages between 1801 and 1803, neither of which survived. After an unsuccessful run to Cornwall, the first of the two was put away in a barn with its boiler still hot while Trevithick and a companion "comforted their hearts with roast goose and proper drink," according to a contemporary account. The carriage burned up while they feasted. According to *Autoweek* of February 15, 1999, however, British steam champion Tom Brogden has built an exact replica of the 1803 London Steam Carriage, complete with three horsepower and a maximum speed of eight miles per hour.
6. Ralph Stein, *The Treasury of the Automobile* (New York: Golden Press, 1961), p. 20.
7. Stephen W. Sears, *The Automobile in America* (New York: American Heritage Publishing Company, 1977), pp. 12–13. Noting contemporary accounts, Sears writes, "How efficient a dredger it was is not stated; but in any event it never became the prototype for a fleet of steam wagons. . . . Evans could not raise the capital to finance his venture, and the Orukter Amphibolos disappeared from history."

8. Nick Georgana, *The American Automobile: A Centenary* (New York: Smithmark Publishers, 1992), p. 11.
9. Levine, *The Car Solution,* p. 21.
10. Ken Purdy, *Kings of the Road* (Boston: Atlantic Monthly Press, 1949), p. 196.
11. Julian Pettifer and Nigel Turner, *Automania* (Boston: Little, Brown and Company, 1984), p. 55.
12. Ibid., p. 47.
13. *Scientific American,* October 6, 1888, p. 215.
14. Michael Brian Schiffer, *Taking Charge: The Electric Automobile in America* (Washington, D.C.: Smithsonian Institute Press, Washington, 1994), pp. 33–34. Schiffer "assumes that ordinary Turks were properly impressed by their Sultan's electric dog cart," which could allegedly make eight miles per hour "on good roads."
15. Ibid., p. 35.
16. James J. Flink, *America Adopts the Automobile, 1895–1910* (Cambridge: MIT Press, 1970), pp. 25–29.
17. Ibid., p. 234.
18. Porsche of North America official history. Further historical context was gleaned from an interview with Karan Tomaselli of Porsche public relations, June 25, 1998.
19. Crabb, *Birth of a Giant,* pp. 168–69.
20. Sears, *The Automobile in America,* p. 39.
21. Michael Frostick, *Advertising and the Motor Car* (London: Lund Humphries, 1970), p. 60.
22. Mark Smith and Naomi Black, *America on Wheels: Tales and Trivia of the Automobile* (New York: William Morrow and Company, 1986), p. 136. The Chandler was advertised under the heading "If You Really Want Your Wife to Drive."
23. Virginia Scharff, *Taking the Wheel: Women and the Coming of the Motor Age* (New York: Free Press, 1991), p. 37.
24. Schiffer, *Taking Charge,* pp. 158–59.
25. *Special Interest Autos* 166 (July–August 1998), pp. 26–27.
26. J. C. Furnas, "Are Electric Cars Coming Back?" *Saturday Evening Post,* March 12, 1960. The story was written just as smog was beginning to become a recognized problem in Los Angeles and other cities. "Many American cities would have less smog trouble if taxis, delivery trucks and a sizable proportion of private automobiles were electric," Furnas wrote. His article also covered early experiments with fuel cells, one of which had already been installed in an Allis-Chalmers tractor. A "Captain Grimes" of Electric Storage Battery was quoted as predicting that "a high-speed, long-range, fuel-cell-

powered electric car" would be in prototype form within three years. It didn't happen.

27. Barbara Whitener, *The Electric Car Book* (Louisville: Love Street Books, 1981).
28. Levine, *The Car Solution,* pp. 123–24. The author also notes the work done by one Wallace J. Minto on a "flourocarbon" (sic) car.
29. Ibid.

Chapter 2 A DIZZYING RIDE: INTERNAL COMBUSTION'S RAPID RISE AND COMING DECLINE

1. Matthew L. Wald, "Number of Cars Is Growing Faster Than Human Population, *New York Times,* September 21, 1997, p. 35. A study conducted by the Department of Transportation revealed that, in the twenty-eight years between 1969 and 1997, the number of cars in the U.S. grew by 144 percent, to 176 million. The driver pool grew by only 72 percent.
2. Steve Nadis and James J. MacKenzie, *Car Trouble: A World Resources Guide* (Boston: Beacon Press, 1993), pp. 9, 27.
3. "The Car Trap," *World Press Review,* December 1996, p. 6.
4. CALSTART, "FastFacts" fax edition, March 1, 1995.
5. World Bank press release, "Combating the Risks of Air Pollution in Asia's Cities," July 27, 1998.
6. Quoted in "Smog from the Middle Kingdom," *Earth Island Journal,* Summer 1998, p. 3.
7. The material on China's budding auto industry is quoted in Gar Smith, "Buick Does Beijing," *Earth Island Journal,* Spring 1997, p. 26.
8. Peter Collier and David Horowitz, *The Fords: An American Epic* (New York: Summit Books, 1987), p. 42.
9. The Editors of *Automobile Quarterly, GM: The First 75 Years of Transportation Products* (Princeton: Automobile Quarterly Publications, 1983), p. 16.
10. Leon Mandel, *American Cars* (New York: Stewart, Tabori & Chang, 1982), pp. 48–54.
11. Crabb, *Birth of a Giant,* p. 161.
12. Collier and Horowitz, *The Fords,* p. 128.
13. David Gelernter, *1939: The Lost World of the Fair* (New York: Free Press, 1995), p. 364. It's interesting to note that Robert Moses, then the nation's preeminent road builder, didn't embrace the transcontinental highways envisioned in Norman Bel Geddes's Futurama ex-

hibit, denouncing the concept as "plain bunk." Moses thought highways should feed cities, not bypass them.

14. Much of the material on what happened to the trolley system is quoted from a PBS *POV* documentary entitled *Taken for a Ride,* coproduced by Jim Klein and Martha Olson. Brad Snell, who is based in Washington and has spent the last two decades writing the definitive corporate biography of GM for Knopf, contributed to their reporting. He says that federal prosecution of the case was sparked by a whistleblower who brought the authorities detailed files on the auto giant's actions. "They had to get rid of the streetcars," Snell says. "They wanted the space that the streetcars used for automobiles."

15. Mark Hertsgaard, *Earth Odyssey: Around the World in Search of Our Environmental Future* (New York: Broadway Books, 1998), p. 107. Hertsgaard notes that "the profits gained from such skulduggery were incalculable (the scam essentially secured monopoly status for the automobile within the U.S. transportation system), but the punishment was next to nothing."

16. Cliff Slater, "General Motors and the Demise of Streetcars," *Transportation Quarterly,* Summer 1997, pp. 45–66. Slater quotes a New York City official of 1920, Grover Whalen, as wholeheartedly campaigning for the municipal bus. "Let me say emphatically that the trolley can be relegated to the limbo of discarded things, along with the stage coach, the horsecar and the cable car," Whalen said. But he would have seen bus exhaust pipes as offering the smell of progress, not as a health hazard.

17. Todd Purdum, "Spring Is in the Los Angeles Air, but the Smog, Chased by El Niño, Is Not," *New York Times,* April 21, 1998, p. A10. On good days in downtown Los Angeles, the San Gabriel Mountains north of the city are now clearly visible.

18. Robert Dawson and Gray Brechin, *Farewell, Promised Land* (Berkeley: University of California Press, 1999), p. 99. One of the dubious advantages of the private car, the authors report, was the opening up of pristine hillsides for development. Streetcars had been unable to make the climb.

19. Mike Davis, *Ecology of Fear: Los Angeles and the Imagination Disaster* (New York: Henry Holt, Metropolitan Books, 1998), p. 62.

20. An excellent capsule chronology of Los Angeles' air-quality history is available on the California Air Resources Board website at http://www.arb.ca.gov/html/brochure/history.htm.

21. Quoted in Ralph Nader, *Unsafe at Any Speed* (New York: Grossman Publishers, 1965), p. 149. A Professor McFarland is cited as the author of the study.

22. California Air Resources Board website.

23. Lynn Graebner, "Electric Cars: Big Oil, Detroit vs. Big Electric," *Sacramento Business Journal,* November 20, 1995, p. 1. Graebner details the interlocking structures of CAUCA and CAHT. Mangels herself has been quoted as saying, "I believe most, if not all, of our funding comes from WSPA—that's no secret."
24. The remarks are included in an AAMA "Request for Proposal," March 24, 1995.
25. Radio talk-show host Lee Rogers's questions, and guest Anita Mangels's answers, are quoted from an Audio Video Reporting Services transcript, September 16, 1996.
26. Western States Petroleum Association fact sheet, "Electric Vehicle Mandates, Subsidies: High Cost for Small Clean Air Benefits." The handout, given to Californians in 1995, cited a Sierra Research study estimating that the total cost of EV and alternative-fuel mandates would be $28 billion over the next fifteen years.
27. The Ford and GM comments on the CARB mandates were delivered to the Consumer Marketability Forum on Electric Vehicles, June 28, 1995.
28. John R. White, "The Electric Car Is Just Around the Corner—Still," *Boston Globe,* January 31, 1998, p. D1. White derides a 1968 American Motors electric concept car called the Amitron, which he said "had all the styling grace of a gaudy little doorstop." What White called "the Green Gestapo" was ignoring the advances in clean internal-combustion engines.
29. The California Public Interest Research Group Charitable Trust, "Pollution Politics II: The Oil and Automobile Industries' Political Expenditures to Influence California Public Policy, 1991–1995," a research study, 1995.
30. *Rachel's Environment and Health Weekly,* "History of Precaution—Parts 1 and 2," March 27, 1997, and April 3, 1997. This is a comprehensive treatment of how "ethyl" became an automotive octane booster and life extender for the more powerful, high-compression automobiles that Detroit wanted to build.
31. Nader, *Unsafe at Any Speed,* p. 153.
32. Ibid., p. 158.
33. Cited in the *Congressional Record,* September 3, 1969, p. 4. To make cars any cleaner than the 1970 models, Heinen said, would cost consumers "billions of dollars."
34. Paul Rauber, "The Little Engine That Couldn't," *Sierra,* July–August 1998, p. 18.
35. Industry comments are collected on the "Chicken Little" site maintained by the Washington-based Environmental Working Group at www.chickenlittle.org.

36. Ibid.
37. Ibid.
38. William K. Stevens, "Earth Temperature in 1998 Is Reported at Record High," *New York Times,* December 18, 1998, p. A32. Stevens quotes Dr. Phillip D. Jones, a climatologist at the University of East Anglia in England, as saying that 1998 is likely to be "not only the warmest year in thermometer record, but also the warmest year of this millennium."
39. L. F. Ivanhoe, "Future World Oil Supplies: There Is a Finite Limit," *World Oil,* October 1995, p. 77. As Ivanhoe notes, "There are indications that most of the large exploration targets have been found, at the same time that the world's population is exploding."
40. The American Automobile Manufacturers Association, "Key Facts About America's Car Companies: Environment," posted on the group's website in 1997.
41. July 24, 1995, testimony before the Subcommittee on Energy and Power, Committee on Commerce, U.S. House of Representatives.
42. Mitch McCullough, "On the Hill," *Autoweek,* November 23, 1998, p. 26.
43. Christopher Cooper, "Odd Reservoir Off Louisiana Prods Oil Experts to Seek a Deeper Meaning," *Wall Street Journal,* April 16, 1999, p. 1. *Scientific American*'s piece "The End of Cheap Oil," by Colin J. Campbell and Jean H. Laherrère, appeared as part of a March 1998 special report entitled "Preventing the Next Oil Crunch."
44. *GM International Newsline,* October 27, 1998.
45. Daniel Yergin, *The Prize: The Epic Quest for Oil, Money and Power* (New York: Simon and Schuster, 1991), pp. 19–28.
46. Colin J. Campbell, "Oil in Crisis: The Dark at the End of the Tunnel," *Energy Investor,* June–July 1998, pp. 4–5.
47. Gregg Easterbrook, "The Coming Oil Crisis—Really," *Los Angeles Times,* June 7, 1998, p. M1.
48. Union of Concerned Scientists, "Emission Benefits of Electric Vehicles in the Northeast," a research study, July 1994, p. 9.
49. Daniel Sperling, "Gearing Up for Electric Cars," *Issues in Science and Technology,* Winter 1994, p. 34.
50. James Cannon, *Harnessing Hydrogen: The Key to Sustainable Transportation* (New York: INFORM Inc., 1995), p. 218.

Chapter 3 ENGINES OF INGENUITY: NEW TECHNOLOGIES FOR THE CLEAN CAR

1. American Council of Learned Societies, *Dictionary of Scientific Biography,* vol. 1 (New York: Charles Scribner's Sons, 1970), pp.

559–60. Grove felt that his contributions to the embryonic field of energy conservation was underappreciated, and he was probably right: Many of his experiments led to important and acclaimed breakthroughs by other scientists.

2. "Fill 'Er Up Fuel Cell," *Discover,* July 1998, p. 103. Jeff Bentley thinks his gasoline reformer can double the mileage of ordinary internal-combustion cars at greatly reduced emissions. The challenge, he says, is to produce a fuel-cell car that people will want to buy.
3. Mariette DiChristina, "What Really Downed the *Hindenburg?*" *Popular Science,* November 1997, p. 71.
4. Schiffer, *Taking Charge,* p. 7.
5. Sheldon R. Shacket, *The Complete Book of Electric Vehicles,* 2nd ed. (Chicago: Domus Books, 1981), p. 17. The author, smitten by the Detroit's apparent usability on today's roads, opines, "Let us hope that we will see some of our 1981 models 'street working' in 60 years or so."
6. Michael Shnayerson, *The Car That Could: The Inside Story of GM's Revolutionary Electric Vehicle* (New York: Random House, 1996), p. 82. The fast-paced design and launch of the EV1, which makes for exciting reading, wasn't matched by the car's performance in the marketplace. In its first three years on the market in California and Arizona, only slightly more than six hundred were leased to customers.
7. Amy Salzhauer, "Plastic Batteries: All Charged Up and Waiting to Go," *Technology Review,* July–August 1998, pp. 60–66.
8. Shacket, *The Complete Book of Electric Vehicles,* pp. 29–30.
9. Ibid., p. 121.
10. Rosen Motors press release, "World's First Successful Road Test of Turbine-Flywheel Powertrain for Automobiles," January 7, 1997. An earlier test, in August 1996, was not so successful. The car just sat at Willow Springs Race Track, outside Los Angeles, for fourteen hours. The Rosen Motors technology has been licensed by another firm and may reappear in another form.
11. Will Hively, "Reinventing the Wheel," *Discover,* August 1996, pp. 58–68. The magazine reports that actor Kevin Costner believes the premise of his film *Waterworld,* that "our dependence on fossil fuels may soon change the climate, catastrophically."
12. Honda press release, "Future Cars Put Environment First—Without Consumer Compromises," December 10, 1997.
13. Associated Press report, "Honda Unveils Ultraclean Gas Engine," October 20, 1997. The analyst was George Peterson, president of AutoPacific Inc. in Santa Ana, California.
14. Data on this is available in a Union of Concerned Scientists study

called "Reducing Emissions from Motor Vehicles in the Real World," available on the group's website at http://www.ucsusa.org/ucs.transportation.html.

15. William J. Cook, "Piston Engine, R.I.P.?" *U.S. News and World Report,* May 11, 1998, p. 46.
16. Union of Concerned Scientists undated report, "Transforming Transportation: Diesel Engines and Public Health," p. 1.
17. Sara Thurin Rollin, "How Hazardous Is Diesel Exhaust?" *SEJ Journal,* Spring 1998. A California scientific panel referenced in the story estimated that 450 out of one million people exposed to diesel exhaust will develop cancer during their lifetimes. Elevated cancer levels have been found in railroad, trucking, and bus garage workers exposed to high levels of diesel fumes.

Chapter 4 Road Warriors and Early Adopters: Living with a Battery-Powered EV

1. Schiffer, *Taking Charge,* p. 74.
2. Policy Resources Group of Schoharie, E. J. McMahon et al., 1998.
3. E-mail from Colin Summers, August 18, 1998.
4. The examples in this paragraph are all quoted in Schacket, *The Complete Book of Electric Vehicles.*
5. Matt Scanlon, "Dave Arthurs' Amazing Hybrid Electric Car," *Mother Earth News,* June–July 1993, pp. 40–75.
6. Shnayerson, *The Car That Could,* p. 247.
7. Jacquelyn Ottman, *Green Marketing: Opportunity for Innovation,* 2nd ed. (Chicago: NTC Business Books, 1998), p. 6.
8. Ibid., p. 3.
9. G. Christian Hall, "Bringing It Home," *Wall Street Journal,* June 16, 1997, p. R1.

Chapter 5 U-Turn: The Big Three Get Serious About Green Cars

1. *Vancouver Sun,* July 30, 1998, p. 34. In reporting on Vancouver's air emergencies, the *Sun* quotes resident Avril McLauchlan, whose two-year-old son has asthma. "The last couple of days have been pretty gross," she says.
2. Joanna Piros, "Knowledge Is Power," *Vancouver Lifestyles,* April 1998, p. 42–48. The boys from Ballard are hometown heroes. Ballard, the magazine notes with pride, "was born in North Vancouver."
3. Jacques Leslie, "Dawn of the Hydrogen Age," *Wired,* October 1997, p. 143.

4. Piros, "Knowledge Is Power," pp. 42–48. After the gala ceremony, somebody went and got the right bolt, then Ballard gave everyone rides in its now fully functional bus.
5. Matthew L. Wald, "Fuel Cell Will Supply All Power to a Test House," *New York Times,* June 17, 1998. Wald, a *Times* transportation writer, has been consistently ahead of the pack in reporting on fuel-cell developments.
6. Kathleen Kerwin, Keith Naughton, and Aaron Bernstein, "GM: It'll Be a Long Road Back," *Business Week,* August 17, 1998, pp. 38–39.
7. Shnayerson, *The Car That Could,* pp. 6–7.
8. Joseph B. White, "GM, Energy Conversion Joining Forces to Develop New Electric Car Battery," *Wall Street Journal,* March 10, 1994, p. B5. The newspaper reports that Stempel joined Energy Conversion's board in January 1994. He resigned that post at the battery maker a few months later, apparently because of conflict-of-interest worries when it appeared that the Energy Conversion–GM deal would go through.
9. Smith and Black, *America on Wheels,* pp. 215–16. The authors leave no stones unturned in their search for automotive oddities.
10. *Automotive Composites: A Design and Manufacturing Guide,* a pamphlet issued by Ray Publishing, 1997. The guide points out that the first fiberglass-bodied car was the 1953 Chevrolet Corvette, and that the 1975 Chevrolet Monza had polyurethane bumpers. Plastic-bodied cars sometimes complicate recycling efforts, since up to sixty different plastics—all with different recycling requirements—can be used in a single car.
11. Alex Taylor III, "Ford's Really Big Leap at the Future," *Fortune,* September 18, 1995, p. 134. Taylor opines that Ford "prospered in the 1950s and 1960s by copying GM in everything from management style to model design."
12. Donald W. Nauss, "Ford Investing $420 Million for Fuel-Cell-Powered Auto," *Los Angeles Times,* December 16, 1997, p. A1.
13. Matthew L. Wald, "Ford Plans Zero-Emission Fuel-Cell Car," *New York Times,* April 22, 1997, D4.
14. C. E. "Sandy" Thomas, "Solar Hydrogen: A Sustainable Energy Option," *Solar Today,* September–October 1993, p. 11. Obviously, you could use those photovoltaic collectors to directly produce electricity. "But solar electricity has three limitations," Thomas writes. "Photovoltaics and wind are intermittent, electricity cannot be economically stored, and solar electricity cannot effectively power motor vehicles."
15. Keith Bradsher, "The Top Spot at Ford Is Returning to a Ford," *New York Times,* September 12, 1998, p. C1. The fear in Ford's tradi-

tional power structure is palpable as Bradsher cites young Ford's "close ties to environmentalists and his periodic calls for the auto industry to pay closer attention to environmental issues."

16. Daniel McGinn, "The New Ford," *Newsweek,* November 23, 1998, p. 54. *Newsweek* quotes historian Douglas Brinkley, who's writing a book about the Ford company, as saying that Bill Ford "has a visionary streak that hasn't been seen since the original Henry Ford. It's not enough for him to run the company—he wants to make the automobile pollution-free."
17. Keith Bradsher, "Making Tons of Money and Fords, Too," *New York Times,* February 14, 1999, p. 14. According to Bradsher, it costs Ford only a few thousand dollars more to make an Expedition than it does the pickup truck on which the sport-utility vehicle is based. But the Expedition sells for as much as $10,000 more.
18. James Bennett, "Chrysler, with Misgivings, Will Sell Electric Mini-Vans," *New York Times,* May 4, 1994, p. D1. While the assembled executives said the EPIC was "their minivan 'state of the art' for electric vehicles," they also "declared that the art was miserable."
19. Transcript dated June 28, 1995. Glaub also said that "a California government regulation cannot create a market for products that people may not want. . . . In other words, mandates do not work in a consumer-driven, free market economy."
20. *An Overview of Hybrid Technology,* Chrysler Corporation pamphlet, 1994. "Efficiencies theoretically inherent in hybrid power-trains are not always realized in practical applications," the pamphlet noted.
21. David Lawder, "Chrysler Develops Gasoline Fuel Cell for Cars," *Reuters Business Report,* January 6, 1997.
22. McGinn, "The New Ford," p. 56. Robert Kennedy Jr., an environmental lawyer, has traded in his GM minivan as "my way of supporting Bill Ford."

Chapter 6 The Global Green Car: Germany and Japan on the Fast Track

1. Cannon, *Harnessing Hydrogen,* p. 99–108. Cannon's book is the best guide to understanding hydrogen infrastructure issues.
2. *This Week with Sam Donaldson and Cokie Roberts,* May 10, 1998. Donaldson pressed Chrysler vice president Robert Liberatore to admit that the "merger" was in fact an acquisition. "It's a merger that's in the interests of Chrysler's workers," Liberatore said. "It will be a joint venture company called DaimlerChrysler." Donaldson

shot back, "The majority of which is going to be owned by foreign investment?"

3. "The World Is Ready for Greener Cars," *Time,* October 21, 1997. The interview was in an advertising supplement that appeared in *Time*'s international editions as part of a special issue on the environment.
4. Alex Taylor III, "How Toyota Defies Gravity," *Fortune,* December 8, 1997, p. 100. Taylor reports that Ford, GM, and Chrysler are regular visitors to Toyota's manufacturing complex in Georgetown, Kentucky. The free, five-hour factory tours for the industry are booked months in advance.
5. *Toyota Alternative Fuel Vehicles Vision,* Spring 1998, p. 1.
6. Center for Resource Solutions press release, "Toyota Motor Sales USA Becomes the First Green-*e* Certified Company," May 8, 1998.
7. CALSTART press release, "Toyota Receives CALSTART's Top 1997 Blue Sky Award for Advanced Transportation Leadership," December 29, 1997. Toyota, said CALSTART president Michael Gage, is "currently setting the trend globally in clean, efficient vehicles."
8. *Electric Vehicle Today,* a daily one-page faxed news summary, July 24, 1998.
9. Dave Phillips, "Honda to Share Clean-Air Technology, Fuel Research," *Detroit News,* February 26, 1998.
10. Jon Pratty, "Cabbies Get Chance to Clean Up Their Act," *Daily Telegraph,* July 2, 1998.
11. *Hydrogen & Fuel Cell Letter,* "ZEVCO Unveils Fuel Cell Taxi, Shell UK Chief Says Company Is into Hydrogen for Real," August 1998.
12. Ibid.
13. Roland Gribben, "BP to Withhold Part of API Dues," *Daily Telegraph,* May 26, 1998.
14. Transcript provided by PSA, dated December 2, 1995. Helmer compared EVs to microwave ovens, whose sales initially stagnated in the marketplace after achieving 10 percent penetration.
15. Audrey Choi and Gabriella Stern, "The Lessons of Ruegen: Electric Cars Are Slow, Temperamental and Exasperating," *Wall Street Journal,* March 30, 1995, pp. B1–B2. The authors recount the sad experience of one Richard Stemmler, who couldn't get a charge in Binz but decided to try to get home anyway. "The car stalled just before the final hill, less than a mile from his house. Reluctant to wake up his father, he walked home, borrowed his father's gasoline-powered car, drove back to Binz, hauled a friend out of bed and returned to the stranded electric BMW. Together, they towed the car home and plugged it in."

16. Rob Edwards, "Does Greenpeace Have Designs on the Motor Industry?" *New Scientist,* November 1995, p. 9. German car industry officials scoffed at Greenpeace's car, claiming it may not be safe to drive.

Chapter 7 THINKING ABOUT TOMORROW: VISIONARIES, PESSIMISTS, AND INVESTORS AT THE CROSSROADS

1. Amory B. Lovins and L. Hunter Lovins, "Reinventing the Wheels," *Atlantic Monthly,* January 1995, pp. 75–86.
2. Amory B. Lovins, John W. Barnett, and L. Hunter Lovins, "Supercars: The Coming Light-Vehicle Revolution," a Rocky Mountain Institute technical paper delivered in May–June 1993.
3. C. E. "Sandy" Thomas, Brian D. James, and Franklin D. Lomax Jr., "Market Penetration Scenarios for Fuel Cell Vehicles," a research paper from Directed Technologies, Inc., 1997.
4. Joan M. Ogden, "Developing an Infrastructure for Hydrogen Vehicles: A Southern California Case Study," a research paper from the Center for Energy and Environmental Studies, Princeton University, 1998.
5. P. J. O'Rourke, "Dog-Faced Bureaucrats Hate Trucks," *Automobile,* July 1998, pp. 52–53. The column is called "This Space for Rant," and O'Rourke does. "The regulatory zombies are lurching across the fields of public discourse wrapped in the ghastly shrouds of 'safety,' 'environmentalism' and 'consumer advocacy,' " he writes.
6. Nadis and MacKenzie, *Car Trouble,* p. 27.
7. Brock Yates, "Fuel-Cell Miracles and Urban Sprawl," *Car and Driver,* August 1999, p. 30. In the column, Yates also opined that building more roads actually reduces pollution, because travel becomes faster and more efficient. This may come as news to the suburban commuters struggling through gridlock around most of our major cities.
8. E-mail from Robert Cumberford, November 26, 1998.
9. Joe Lorio, Georg Kacher, Paul Lienert, Richard Feast, and Mark Gillies, "Double Vision in Tokyo," *Automobile,* January 1998, pp. 62–69. A year later, covering the Detroit Auto Show, *Autoweek* for January 18, 1999, ignored the Honda VV hybrid prototype and pronounced a Chrysler sport-utility vehicle the "most significant" in the show.
10. Steve Lohr, "Long Boom or Bust: A Leading Futurist Risks His Reputation with Ideas on Growth and High Technology," *New York Times,* June 1, 1998, pp. D1, D6.

11. Carolyn Leitch, "Ballard Blasts to New High," Toronto *Globe and Mail,* March 18, 1998, p. B1.
12. Nadis and MacKenzie, *Car Trouble,* p. 23.
13. Ibid., p. 25.
14. Ralph Nader and William Taylor, *The Big Boys: Power and Position in American Business* (New York: Pantheon Books, 1986), p. 71. DeLorean was also accused of showing, as Nader and Taylor put it, "a distracting concern over minority rights, urban poverty, and other social issues."
15. Ibid., p. 75. In his amazingly frank memo about GM, DeLorean pointed out that, because of the company's dominant place in the market, "we also, in effect, control our competitors—who would be economically devastated if they tried to do better socially but at greater product cost."

Chapter 8 Jump-Starting the EV: Federal Funding for Alternative Fuel

1. Sierra Club press release, "EPA Clean Air Plan Earns 2.5 Cheers from the Sierra Club," February 18, 1999. SUVs of more than six thousand pounds, including the Ford Excursion, would be subject to lighter smog standards under the proposals.
2. U.S. Public Interest Research Group press release, "Study Finds New SUV Rules Could Cut over 1.2 Million Tons of Smog-Forming Pollution," March 17, 1999. Light trucks, the report says, "are allowed to emit up to three times as much nitrogen oxide as passenger cars."
3. Matthew L. Wald, "Junked Vehicles Outperform New Models," *New York Times,* August 11, 1997, p. D1.
4. Nadis and MacKenzie, *Car Trouble,* p. 50.
5. Ibid., p. 148.
6. Jane Holtz Kay, *Asphalt Nation* (New York: Crown Publishers, 1997), pp. 347–48.
7. Kenneth Cole, "Big Three, Fed Allies Narrow Options," *Detroit News,* July 8, 1998.
8. National Research Council, *Review of the Research Program of the Partnership for a New Generation of Vehicles, Third Report* (Washington: National Academy Press, 1997), p. 3.
9. Morton Mintz, "Smog Fighter Inspired Auto Industry Lawsuit," *Washington Post,* January 26, 1969.
10. Ibid.
11. Automobile Manufacturers Association press release, January 10, 1969. In a counteroffensive, the AMA said the industry's signifi-

cant accomplishments were made possible only by "the cross-fertilization of ideas and the full exchange of technical information among automobile manufacturers and suppliers."

12. Cannon, *Harnessing Hydrogen,* pp. 294–95.

Chapter 9 CLEARING THE AIR: CLEAN CARS AND SUSTAINABLE TRANSPORTATION IN THE TWENTY-FIRST CENTURY

1. Fritz R. Kalhammer, Paul R. Prokopius, Vernon P. Roan, and Gerald E. Voecks, *Status and Prospects of Fuel Cells as Automobile Engines: A Report of the Fuel Cell Technical Advisory Panel,* a California Air Resources Board report released in 1998, pp. iv–1. This study is the most up to date and comprehensive look at the possibilities for fuel-cell cars in the twenty-first century. Its focus is not local but global, and more proof of CARB's international influence.
2. Ibid., p. 5. The panel concludes that the optimistic public statements about fuel-cell cars are "bona fide expressions of the automakers' confidence in their plans."
3. P. Chen, X. Wu, J. Lin, and K. L. Tan, "High H2 Uptake by Alkali-Doped Carbon Nanotubes Under Ambient Pressure and Moderate Temperatures," *Science,* July 2, 1999.
4. Ogden, "Developing an Infrastructure for Hydrogen Vehicles," pp. 16–17. Ogden's sober analysis of renewable hydrogen's potential deserves a wider audience. And a report looking at this option globally would be very useful.
5. Clean Air Trust press release, "Sulfur in Gasoline: The Real Story," March 19, 1998. Initially, the oil companies said that removing sulfur from gasoline would cost as much as twenty cents a gallon.
6. Donald Nauss, "Putting Autos on a Steel Diet," *Los Angeles Times,* August 11, 1997. Another carmaker trying to reduce the weight of the car's steel body is Germany's Porsche, which estimates its innovations could shave $150 off the cost of a typical $1,100 five-passenger frame.
7. "The Future of SUVs: A New Design Breaks the Mold to Make a Better Sport-Utility Vehicle," *Consumer Reports,* September 1998, p. 13.
8. The Friends of the Earth "Roadhog Info Trough" website, full of eye-opening information about sport-utilities, is at http://www.suv.org. The environmental group is trying to catch the attention of web-surfing car buyers by advertising on automobile-oriented sites and search engines.
9. Cited by Business Wire, September 4, 1998.

10. Cal Thomas, "Beware: Big Government Is Trying to Regulate Us into Riskier Vehicles," *The Connecticut Post,* February 17, 1998, p. A10. Thomas might be interested in a July 15, 1999, *New York Times* story by Keith Bradsher, on page A21, noting that SUV occupants "have roughly the same chance as car occupants of dying in a crash."
11. The statistics in this paragraph are quoted in Allen R. Myerson, "U.S. Splurging on Energy After Falling Off Its Diet," *New York Times,* October 22, 1998, p. C6.
12. Nadis and MacKenzie, *Car Trouble,* p. 118.
13. Myerson, *New York Times,* p. C6.
14. Alan Thein Durning, *The Car and the City* (Seattle: Northwest Environment Watch, 1996), p. 7.
15. Ibid.
16. Quoted in Wolfgang Zuckerman, *End of the Road: From World Car Crisis to Sustainable Transportation* (Post Mills, Vt.: Chelsea Green, 1991), p. 272.
17. Ibid., p. 262.

SELECT BIBLIOGRAPHY

Alvord, Katie. *Divorce Your Car! . . . And Live Happily Ever After.* Gabriola Island, British Columbia: New Society Publishers, 2000.

Angelucci, Enzo, and Alberto Bellucci. *The Automobile from Steam to Gasoline.* New York: McGraw-Hill, 1974.

Brant, Bob. *Build Your Own Electric Vehicle.* Blue Ridge Summit, Penn.: TAB Books, 1994.

Christianson, Gale E. *Greenhouse: The 200-Year Story of Global Warming.* New York: Walker and Company, 1999.

Collier, Peter, and David Horowitz. *The Fords: An American Epic.* New York: Summit Books, 1987.

Crabb, Richard. *Birth of a Giant: The Men and Incidents That Gave America the Motorcar.* Philadelphia: Chilton Book Co., 1969.

Davis, Mike. *Ecology of Fear: Los Angeles and the Imagination of Disaster.* New York: Metropolitan Books, 1998.

Durant, John. *Predictions.* New York: A. S. Barnes and Company, 1956.

Durning, Alan Thein. *The Car and the City.* Seattle: Northwest Environment Watch, 1996.

The Editors of *Automobile Quarterly. GM: The First 75 Years of Transportation Products.* Princeton: Automobile Quarterly Publications, 1983.

Engwicht, David. *Reclaiming Our Cities and Towns: Better Living with Less Traffic.* Philadelphia: New Society Publishers, 1993.

Flink, James J. *America Adopts the Automobile, 1895–1910.* Cambridge: MIT Press, 1970.

Gelernter, David. *1939: The Lost World of the Fair.* New York: Free Press, 1995.

Georgana, Nick. *The American Automobile: A Centenary.* New York: Smithmark Publishers, 1992.

Hertsgaard, Mark. *Earth Odyssey: Around the World in Search of Our Environmental Future.* New York: Broadway Books, 1998.

Ingrassia, Paul, and Joseph B. White. *Comeback: The Fall and Rise of the American Automobile Industry.* New York: Simon and Schuster, 1994.

Kay, Jane Holtz. *Asphalt Nation.* New York: Crown Publishers, 1997.

Latham, Caroline, and David Agresta. *Dodge Dynasty: The Car and the Family That Rocked Detroit.* San Diego: Harcourt Brace Jovanovich, 1989.

Levine, Gary. *The Car Solution: The Steam Engine Comes of Age.* New York: Horizon Press, 1974.

Mandel, Leon. *American Cars.* New York: Stewart, Tabori & Chang, 1982.

McCrea, Steve. *Why Wait for Detroit? Drive an Electric Car Today.* Fort Lauderdale: South Florida Electric Auto Association, 1992.

Montagu of Beaulieu, Lord, and Anthony Bird. *Steam Cars: 1770 to 1970.* New York: St. Martin's Press, 1971.

Nader, Ralph. *Unsafe at Any Speed.* New York: Grossman Publishers, 1965. An expanded version of the book was issued by Knightsbridge Publishing in 1991.

Nader, Ralph, and William Taylor. *The Big Boys: Power and Position in American Business.* New York: Pantheon Books, 1986.

Nadis, Steve, and James J. MacKenzie. *Car Trouble: A World Resources Guide.* Boston: Beacon Press, 1993.

Perrin, Noel. *Solo: Life with an Electric Car.* New York: W. W. Norton, 1992.

Pettifer, Julian, and Nigel Turner. *Automania.* Boston: Little, Brown and Company, 1984.

Purdy, Ken. *Kings of the Road.* Boston: Atlantic Monthly Press, 1949.

Rajan, Sudhir Chella. *The Enigma of Automobility: Democratic Politics and Pollution Control.* Pittsburgh: University of Pittsburgh Press, 1996.

Scharff, Virginia. *Taking the Wheel: Women and the Coming of the Motor Age.* New York: Free Press, 1991.

Schiffer, Michael Brian. *Taking Charge: The Electric Automobile in America.* Washington, D.C.: Smithsonian Institute Press, 1994.

Sears, Stephen W. *The Automobile in America.* New York: American Heritage Publishing Company, 1977.

Shacket, Sheldon R. *The Complete Book of Electric Vehicles,* 2nd ed. Chicago: Domus Books, 1981.

Sherman, Joe. *Charging Ahead.* New York: Oxford University Press, 1998.

Shnayerson, Michael. *The Car That Could: The Inside Story of GM's Revolutionary Electric Vehicle.* New York: Random House, 1996.

Sperling, Daniel. *Future Drive: Electric Vehicles and Sustainable Transportation.* Washington, D.C.: Island Press, 1995.

Yergin, Daniel. *The Prize: The Epic Quest for Oil, Money and Power.* New York: Simon and Schuster, 1991.

Zuckerman, Wolfgang. *End of the Road: From World Car Crisis to Sustainable Transportation.* Post Mills, Vt.: Chelsea Green, 1991.

INDEX

ABOUT THE AUTHOR

JIM MOTAVALLI is the editor of *E: The Environmental Magazine* and a journalist who has written for *The New York Times,* the *Los Angeles Times* Syndicate, *Salon, The Guardian,* and many other publications. His writing on population issues won a 1999 Global Media Award from the Population Institute. Mr. Motavalli also hosts a public-affairs radio show and teaches journalism at Fairfield University. He lives with his wife and their two daughters in Fairfield, Connecticut.